Controlling Common Property
Regulating Canada's
East Coast Fishery

In this study of the Newfoundland inshore fishery, David Ralph Matthews sets out to discover how in the past two decades the harvesting and processing of fish have been transformed by changed government policy and by technological advance. He finds that not only the work of the fishermen but also the social and economic life of their communities has been altered.

In his analysis of the nature of property relations governing common-property resources, Matthews contrasts what resources mean for those who make their living from them, and what they mean for those who regulate them. He uses fisheries department and other documents to show how fisheries policy for eastern Canada's inshore fishery changed in the early 1960s, when a focus on the biological conservation of fish stocks gave way to a concern with the social dynamics of property regulation. He draws directly upon interviews, conducted in five fishing villages, that offer rich insights into local perceptions of conditions and practices. The fishing communities used to provide their own regulation; conflict occurred when government views of the nature of resource property regulation, based on assumptions different from those of the local people, were imposed.

In 1991 the Newfoundland inshore cod fishery virtually collapsed. This book looks at the reasons for the collapse. It explores the effect of underlying assumptions in resource policy on environmental change and resource development, and is a valuable case study in the nature of work relations, economic development, and community social psychology.

DAVID RALPH MATTHEWS is Associate Dean, Graduate Studies, McMaster University. He is author of *The Creation of Regional Dependency.*

Controlling Common Property

Regulating Canada's
East Coast Fishery

David Ralph Matthews

UNIVERSITY OF TORONTO PRESS
Toronto Buffalo London

© University of Toronto Press Incorporated 1993
Toronto Buffalo London
Printed in Canada

ISBN 0-8020-2932-9 (cloth)
ISBN 0-8020-7757-9 (paper)

Printed on acid-free paper

Canadian Cataloguing in Publication Data

Matthews, Ralph, 1943–
    Controlling common property: regulating Canada's East Coast fishery

    Includes index.
    ISBN 0-8020-2932-9 (bound) ISBN 0-8020-7757-9 (pbk.)

    1. Fisheries – Newfoundland.   2. Fisheries – Economic aspects –
    Newfoundland.   3. Fisheries – Newfoundland – State supervision.
    4. Fisheries – Newfoundland – Interviews.   I. Title.

    SH224.M38C77 1993 338.3'727'09718 C93-093870-4

To my children,
Adam and Leah Matthews

Contents

x Contents

Acknowledgments

This book combines a series of community studies and a conceptual analysis of the nature of property relations governing common-property resources. The community-based research was made possible by a grant from the Institute of Social and Economic Research at Memorial University of Newfoundland. I wish to thank Douglas House and Peter Sinclair of Memorial University for helping me to get this grant. I also thank the Labour Studies Program at McMaster University for providing me with a small additional grant, which helped fund coding of the interview data.

I conducted approximately half the interviews with fishermen myself. Most of the remainder were carried out by John Phyne, now of St Francis Xavier University but at that time a doctoral student at McMaster University. John not only established rapport with rural fishermen but took a personal interest in this project. In addition, his doctoral thesis, dealing with the role of fishery officers, contributed to my understanding of the intersection between state policy and community action.

I also owe a debt to Eric Dunne, of the federal Department of Fisheries and Oceans, St John's, who willing broke down bureaucratic barriers so that I might obtain the necessary lists of fishermen, from which I was able to select a random sample. Finally, I owe an obvious debt to the many fishermen in the five communities discussed in this book who so willingly agreed to be interviewed in what invariably seemed to be the week that the fish had 'struck in.' The fact that I did not receive a single refusal stands witness to their generosity and to the legendary hospitality of rural Newfoundland.

My interest in and understanding of common-property relations owe much to a fortuitous invitation I received to participate in a workshop on natural rights and property relations sponsored by the Liberty Fund of the United States, which took place at Niagara-on-the-Lake, Ontario, in April 1985. The organizer of that event was Mark Sproule-Jones, a political science colleague at McMaster University who invited me to be one of the participants. While I had conducted research on the Newfoundland fishery prior to that workshop, I had never previously conceptualized the problems facing the fishery in terms of property relations and 'public choice' issues. Mark's kind invitation opened up a new perspective to me. I thank Mark (and, indirectly, the Liberty Fund) most sincerely for providing me with an intellectually stimulating opportunity.

I have benefited from the comments and critiques of a number of colleagues who have read portions of the manuscript. In particular, I thank Patricia Marchak for her extensive comments on the conceptual framework contained in chapter 3, which, together with her own work on property relations in the fishery, were of particular benefit to me in helping me to shape my ideas. Fikret Berkes, Mark Sproule-Jones, and David Tukura also read parts of the manuscript and provided helpful comments. I thank them for their time and their wise insights. I also appreciate the constructive comments provided by the two anonymous readers chosen by the University of Toronto Press. Finally, I have benefited once again from the encouragement of Virgil Duff, Executive Editor, University of Toronto Press. Virgil was a pillar of encouragement when, half way through this manuscript, I needed reassurance in order to carry on, and he has continued to be supportive and encouraging through the final review and editing stages.

However, when all is said and done, writing a book is largely a solitary activity which necessitates that one give up hundreds of hours that one could have spent in other activities. The unwitting 'victims' of that sacrifice are invariably those to whom one is closest – one's family. My wife, Anne Martin Matthews, was working on a book of her own through much of this period and at least could understand the process involved and empathize with it. Nevertheless, the debt I owe her for her support, concern, advice, and love is immeasurable. The other persons most closely affected by the writing of this work are our children, Adam, aged nine, and Leah, aged eight. Their arrival in my life has given it a meaning that I simply could not appreciate was possible beforehand. However, during a significant portion of their young lives I have been working on this volume. Anne and I have attempted to ensure that our additional work activities have not been at the expense of time taken from our children, but there must be some things that I might have done with and for them had I not been engaged on this project. The only compensation that I can give is to dedicate this work to them, with love.

Controlling Common Property
Regulating Canada's
East Coast Fishery

1 Setting Out

An Introduction to the Analysis

There is no such thing as an average fisherman.

Navigating Troubled Waters: A New Policy for the Atlantic Fisheries.
Report of the Task Force on Atlantic Fisheries (Kirby Task Force 1982, 14)

Too often sociologists forget that a major aim of their discipline is the
analysis of persons in varying forms of social organization. An exclusive
focus on aggregately distributed persons restricts such generalizations,
whereas a major aim of any study should be the observation of persons
in natural interactive units ... The investigator must be clear as to the
interactional unit under analysis and take pains to insure that the
appropriate units are observed.

Norman K. Denzin,
The Research Act (1970, 82)

The Transformation of Work and Community

This book can be seen as dealing with three general areas of concern.
First, it is concerned with the role of the state, particularly with the
impact of state regulation on the Newfoundland fishery. Second, it is
concerned with the nature of work in the Newfoundland fishery, particularly with the way in which the nature of work has been 'transformed'
through changes in technology and as a result of state regulation.
Third, it is concerned with the nature of community life in rural Newfoundland, particularly with the way in which rural community life is
being altered by the transformation of work in the fishery and by the
role of the state. A better understanding of these three general areas of

concern can be obtained if the dimensions of each of them are outlined in more detail.

The research on which this book is based grew out of the author's interest in the way state policy was attempting to 'rationalize' the nature of work carried out by those Newfoundlanders who operate largely 'inshore' or 'nearshore' in small and intermediate-sized boats.[1] Much more will be said about such policies later in the book. These policies can be understood as an attempt by the government to transform the fishery from a largely traditionally organized activity operating primarily by underwritten rules into a systematically and formally regulated occupation.

The research was undertaken in an attempt to understand better the nature of and reasons for these policies, and to examine the impact they were having on the actual day-to-day activities of Newfoundland fishermen. In particular, the research was intended to focus on the way in which such regulations affect the rights of fishermen with respect to access to the fishery. The fishery is a common-property resource that has traditionally been open to all who wish to pursue it. Recent government legislation has attempted to alter this common-property character of the Newfoundland inshore and nearshore fishery, transforming it from an open-access resource into one that is accessible only to certain fishermen.

This consideration of the role and impact of state regulation on the open-access character of the inshore fishery involves five issues:

1 The nature of state regulation of the Newfoundland inshore and nearshore fishery, particularly with respect to licensing and the limiting of access to what was previously a common-property resource.

2 The extent to which there is conflict between federal and provincial goals and policies with regard to licensing and the regulation of the Newfoundland inshore and nearshore fishery.

3 The way in which those most affected by state regulation of the Newfoundland inshore and nearshore fishery (that is, the fishermen themselves) react to those external measures to control their means of making a livelihood.

4 The extent to which there are local or customary ways in which inshore and nearshore fishermen have traditionally regulated themselves and controlled access to the fishing ground.

5 The relationship between formal (state) regulation of the inshore and nearshore fishery and the informal (customary) regulation of it by the fishermen themselves.

This book also grew out of the author's interest in the way in which the nature of work in the Newfoundland inshore and nearshore fishery was being altered and transformed. Some of the changes were obviously due to changes in the types of boats and gear now being used by some fishermen. Those will be referred to here as different technologies of production.[2] In examining the impact of such technologies on the nature of work, this book will deal primarily with the following concerns:

1 The extent to which there are different technologies of production in the Newfoundland inshore and nearshore fishery.
2 The extent to which there is competition and conflict among fishermen using different technologies of production.
3 The impact of such differences in technology on the nature of work and the organization of community life in rural Newfoundland.

Not all the changes in the nature of work in the Newfoundland fishery can be attributed to changes in the nature of the technology employed. Quite clearly, some of the 'transformations' of work stem directly from the recent state policies developed to regulate the fishery. Thus a fourth concern about the transformation of work is:

4 The way in which state regulation with regard to licensing and the access to the fishery differentially affects those using different technologies, thereby altering the relationships among fishermen using different types of gear and equipment.

Fishermen do not fish only from individual boats; it is fair to say that they also fish from communities. This is perhaps what most distinguishes the world of work in rural communities from the types of industrial and factory work found in larger centres. In urban areas, it may make some sense to have an industrial sociology or a sociology of work that pays scant attention to the broader concerns of community and family life, though recent developments in family sociology and gerontology have called that approach into question. In agricultural and fishing communities, however, the separation of work from community and family life makes no sense and is generally impossible. In Newfoundland, as in fishing and agricultural communities virtually everywhere, it is next to impossible to study the nature of work without being involved, ipso facto, in a study of community and family life. Put slightly differently, in rural Newfoundland, community differences stem largely from the different ways in which fishermen in different communities prosecute the fishery.

Partly because of the relationship between work and community, this study was begun on the assumption that the nature of the inshore and nearshore fishing industry could not be understood without reference to the types of community social organization in which the fishery is embedded. More will be said about this later, as it is a key dimension of the research. But it is worth emphasizing here that the state regulation of the Newfoundland fishery affects not only the work activities of fishermen, but also many other aspects of their general community life. That is, state attempts to 'rationalize' the fishery may alter the social and economic character of the communities in which the inshore fishery is based. That concern prompts a consideration of three issues:

1 The extent to which Newfoundland fishing communities differ from one another and the extent to which such differences are related to the different organization of fishing in those communities.

2 The extent to which there are different communal economies or 'work communities' within the Newfoundland inshore fishery.

3 The way in which licensing and other aspects of the state regulation of the inshore fishery will, in turn, have a unique effect on each of the distinct ways of life found in these communities.

Though the various aspects of state regulation and the transformation of work and community just outlined may be analytically separable, in reality the three dimensions are inextricably related. The goal of this book is to examine and explain their interrelationship.

An Overview of the Book

In the following pages we examine the nature of the fishery as a common-property resource. The philosophical and theoretical antecedents of that perspective will be examined in detail in chapters 3 and 4. It will suffice here, by way of orientation, to explain that the perspective owes its recent popularity largely to the work of Garrett Hardin, who in 1968 vividly depicted what he described as 'the tragedy of the commons.' It was Hardin's argument that any resource that remains the 'common property' of all, without any mechanism for controlling who may have access to that resource, will inevitably be depleted. He based his reasoning on a combination of simple social psychology and basic strategizing. Thus, it was Hardin's contention that in any open-access common-property situation, the 'rational choice' of all harvesters is to secure as much of the resource as possible in the quickest possible time. Should they not do so, and attempt, for example, to conserve the resource so that it might replenish itself, others will take everything that is currently available and thus make such husbandry impossible.

In situations of common property, Hardin argued, to fail to exploit the resource unscrupulously is to set oneself up as a victim.

Hardin's perspective was not particularly new; the essential elements of it are found in the work of a variety of thinkers, including Malthus, Darwin, Adam Smith, and more recent neoclassical economists. Even its application to the fishery was foreshadowed in the work of the Canadian economist Scott Gordon (1954). None the less, Hardin's perspective became the basis of a growing body of thought and writing, particularly in economics and political science. Though that body of work dealt with a wide variety of common-property situations, one of its primary focuses was on the fishery. Ocean fisheries in particular appear to represent almost the archetypal or ideal typical situation that Hardin described. Fish swim freely and without being owned until they are caught and landed in the boat. Only then can they be described as in any sense private property. Moreover, in most areas of the world the right to go in search of fish is largely uncontrolled and is a prime example of an open-access situation. In short, the fishery appears to embody the two attributes (that is, common property and open access) that Hardin contended inevitably result in the complete depletion of a resource.

This book is written on two levels and, to some degree at least, in two parts. Chapters 2, 3, and 4 essentially adopt a broader, or 'macro,' perspective, in which the general issues pertaining to the Newfoundland fishery are examined. They involve a consideration of the historical development of Newfoundland's inshore fishery and the role of state regulation in its development. The underlying leitmotif is an argument that the incorporation of the 'tragedy of the commons' perspective into state policy had a profound effect on the way in which the fishery came to be regulated.

The context for this discussion is provided in chapter 2, which outlines the historical development of Newfoundland in general and of the Newfoundland fishery in particular. It demonstrates that the past two decades have seen a transformation of both the harvesting and the processing sides of the fishery. The transformation is seen to be the result of changing state policies combined with the incorporation of significantly different technologies of production in both the harvesting and the processing sectors. Chapter 3 contains an examination of the shift in value orientation that occurred in Canadian fisheries policy between approximately 1965 and 1982. This chapter documents the growing acceptance of the 'tragedy of the commons' perspective by

Canadian fisheries policy makers and demonstrates the extent to which federal acceptance of this approach resulted in a major conflict of values between federal politicians and officials and their Newfoundland counterparts. In chapter 4 the general theoretical issues surrounding resource property rights are examined. The analytic differences and relationships among a variety of types of property – including state property, private property, and common property – are examined. This chapter also lays the groundwork for a consideration of 'community common property' as analytically distinct from the other types generally discussed in the literature on property relations. Of greatest relevance, however, is the essentially social psychological or phenomenological perspective on property that is developed in this chapter. From this perspective, property and property relations are seen as subject to varying contextual meanings. Our consideration of the property relations that exist in the various Newfoundland fishing communities examined in subsequent chapters is based on such an approach.

Chapters 5, 6, and 7 involve micro-level analyses of the nature of the fishery in five Newfoundland communities of varying size and complexity. These community studies are not simply descriptions of the way in which each type of community conducts its fishing. They are designed, rather, to provide actual empirical settings in which the issues raised in the preliminary chapters may be examined in less abstract and more vivid terms. Although there are many themes woven into these chapters, the most fundamental of all is that the fishery, prior to the introduction of state regulation, was far from the unregulated commons depicted by Hardin and his followers. These community studies demonstrate the rich and wide variety of regulatory practices that existed in each of the five communities, virtually all of them aimed at ensuring that the 'tragedy' of overfishing did not occur. Moreover, the scope of the regulatory practices in most of the communities studied extended to ensuring that the technological transformation of the 1960s and 1970s in both harvesting and processing did not undermine the carefully developed ecological balance that those historic regulatory practices had achieved. Consequently, the other underlying theme of these community studies is the extent to which federal and provincial regulations designed to 'protect' the inshore fishery from overexploitation were incorporated into already-existing community fishing regulations and practices.

The 1980s and 1990s have seen an enormous outpouring of scholarly activity from several disciplines regarding the validity of the 'tragedy of the commons' perspective. Much of that work has demonstrated the existence of traditional regulatory activities in situations where open access was previously thought to exist. The present work can be seen as a contribution to that body of literature. Perhaps more than the earlier studies, however, this work demonstrates what occurs when traditional community-based regulation and centralized state-oriented regulation of the same natural resource come into contact with each other. An additional feature of this work is its examination of these two forms of regulation in five different community settings. This feature permits us to explore in a comparative context the factors that give rise to differing forms of regulatory behaviour with respect to community-based natural resources.

Chapter 8, the conclusion of this work, attempts to do more than provide the usual summary. It develops two themes. First, it explores the actual 'need' for external regulation of the inshore fishery, in view of the fact that the empirical studies contained herein clearly demonstrate the existence of community-based regulatory practices prior to major state attempts to 'police the commons.' This exploration involves a broader consideration, namely, of the place of the inshore community-based fishery in the context of the larger, global prosecution of the North Atlantic fishery. Second, the chapter examines a related issue – that is, the future of the Newfoundland inshore fishery in the face of its recent complete collapse and the decision of the Canadian state to ban most inshore fishing activity in Newfoundland for a period of nearly two years. These developments will be examined in the context of the information and analysis contained in this work.

The Relevance of Previous Research on Newfoundland Fishing Communities

Before describing the research design of the present study, we will review briefly two earlier bodies of work and their influence on it. One significant body of research was the series of community studies carried out under the sponsorship of the Institute of Social and Economic Research at Memorial University of Newfoundland. The other was an extensive analysis of the Canadian East Coast fishery that served as the basis of the report produced by a Canadian government task force (Kirby Task Force 1982).

The Institute of Social and Economic Research (ISER) at Memorial University of Newfoundland was begun in the 1960s, and it quickly established a reputation for funding and publishing anthropological and sociological studies on the nature of social, economic, and political life in rural Newfoundland communities (see Philbrook 1965, Szwed 1966, Faris 1966, Finestone 1967, Iverson and Matthews 1968, Brox 1972, Chiaramonte 1971). However, this early interest did not carry over into the next decade, and, with the exception of work reported in the edited collections by Andersen and Wadel (1972) and Andersen (1979), the development of the inshore fishery during the 1970s went largely unreported and unanalysed. Yet, the 1970s saw considerable change in all sectors of the Newfoundland fishery. As a consequence, in the 1980s ISER renewed its interest in the fishery and funded both a study by Sinclair (1985, 1987) and the community research that is reported here.

These studies were seen as 'following up' the studies of the 1960s and were intended to update our knowledge of the changes in New-foundland fishing communities since the 1960s. Despite this, they differ considerably from their predecessors. First, there is a difference of scope. The earlier studies tended to be general anthropological analyses of all aspects of family and community life. The more recent studies have focused primarily on the organization of fishery labour, and only indirectly on community life. There is also a significant difference in conceptual orientation. The earlier studies were largely descriptive analyses, written without reference to an overarching theoretical perspective. However, one of the major intellectual developments in social science during the 1970s and 1980s was the rise of a political economy perspective that focused on the extent to which state policies were the outcome of class interests. That perspective was an integral part of Sinclair's analysis. Similarly, the present study was designed with the intention of examining the role of the state and the impact of its policies on fishing activity. Only after the empirical data were collected did this writer begin to reconceptualize his findings in the context of the 'tragedy of the commons' perspective.

The second body of research that had a significant impact on the design of this study was a survey carried out in the early 1980s by a federal government commission of inquiry established to investigate the almost complete financial collapse of the Atlantic fishery. As one of its activities, the task force undertook a survey of inshore fishermen throughout the Atlantic provinces, during the course of which some one thousand of approximately twenty thousand inshore fishermen were interviewed. The interviews were carried out primarily by fishery officers, and focused particularly on all sources of income, in cash and in kind, received by the respondents. There were few questions dealing with any other aspect of work or communal life. The data obtained from that study were used to support arguments for the restructuring of the fishery (Kirby Task Force 1982) that ultimately became the basis of state policy.

There were obvious design problems with the task force study. Fishery officers are responsible for policing infractions of fishery regulations, so it was unlikely that they would be given accurate information on the extent to which fishermen deliberately overfished or received income through violations of licensing policies. Similarly, the lack of attention to social data in the interview schedule that was used clearly limited the information obtained about the occupational organization and communal base of fishery.

The other problem with the task force survey was more subtle. While the fishery is an occupation, it is also a community-based way of life, to the extent that the social organization of the fishery differs remarkably even between communities only a few miles apart. Thus an understanding of the communal context of the fishery is important to an understanding of it as an occupation. Given this, the task force study did not employ an adequate methodology, for even a sample as large as a thousand would still contain approximately only one fisherman for every two communities in Atlantic Canada. Although such a survey might be adequate to tell us something of the income patterns of East Coast fishermen, it is not adequate to tell us very much about the fishery as a community-based work activity.

The methodological point that is being made here deals with the issues surrounding the most appropriate unit of analysis in a study of the Canadian East Coast inshore fishery. This sort of argument is certainly not new. As long ago as 1950, Robinson critiqued the 'ecological fallacy' of using aggregate data to explain individual actions (Robinson 1950). From a different perspective, interactionists have argued that a random-sampling procedure 'forces a treatment of society as if society were only an aggregation of disparate individuals' (Blumer 1948, 546).

In this instance we are making a case similar to that of the interactionists by arguing that survey data do not adequately capture the complexity of social life, while supporting the position that the aggregation of individual data cannot explain the nature of communal life. That is, we are arguing that it is the community, not the individual, that is the most appropriate unit of analysis when it comes to understanding the fishery. Because of the community-based character of inshore fishery work, a random sample of all fishermen that picks only one or two persons from each community is inadequate as a methodology. It is inadequate because the information obtained cannot tell us enough about the various ways in which that occupation is organized and conducted. The quotation from Denzin's work at the beginning of this chapter focuses directly on this point. In Denzin's terms, for an understanding of inshore fishery, the 'natural interactive unit' for study is the community. Instead of focusing on how widespread particular attitudes, values, or practices may be, the methodological goal is rather to understand behaviour in the context of the primary social organization involved – which in this case is the work community. The inshore fishery, we are arguing, is one of those instances where systematic survey research is not as appropriate or effective a methodology as a more intensive form of communal analysis.

It is this perspective that guided the research design of this study. For the reasons just cited, the research reported here is based not on a random sample of inshore fishermen, but on a series of community studies. Five Newfoundland communities were chosen as the focus of the research. No claims are being made here that the five communities accurately reflect *all* Newfoundland inshore fishermen. They were chosen, rather, in the hope of demonstrating that a variety of different occupational and communal lifestyles exist within the Newfoundland inshore fishery. If we had studied more communities, we might have discovered even more such lifestyles – or at least considerable variations on the ones we did investigate. Furthermore, because of the small size of most of the communities studied and the even smaller size of our sample of respondents from each, it is difficult to make overall generalizations about the nature of fishing as an activity. Rather, the purpose behind this research was to demonstrate the rich diversity of the community and work contexts of fishing as a way of life, and to establish that fishing as an activity can be understood only in its community context. It is also clear from our data that there are wide variations in occupational lifestyle within most of the communities examined. The statement 'There is no such thing as an average fisherman,' quoted at the beginning of this chapter, was an earlier attempt to make the point.

This discussion was intended to explain how considerations of occupational and community structure guided our research design just as much as considerations of state regulation will guide our ensuing analysis. The specific reasons for choosing each community are discussed below.

Criteria for Choosing Communities and Fishermen for Study

There were two levels of sample selections involved in this study, the selection of communities to be studied and the selection of fishers in each of the communities as the persons to be interviewed.[3] Turning first to the choice of communities, the simple limitations of time and resources made it impossible for us to study a large number of communities. But if only a small number were to be studied, it was important that they be chosen with attention given in advance to differences among them. A random sample of communities was out of the question for reasons of cost and time. Thus, the communities chosen should allow us to investigate the widest possible range of differences in work and community organization.

Three criteria were developed to guide the choice of communities. As the focus of this research was on the inshore fishery, our first criterion was that each community selected for study should contain a predominance of inshore rather than deep-sea fishermen. However, as previously noted, within the inshore fishery there is a fundamental difference between those who engage in the traditional small-boat fishery and those who employ the considerably larger longliners. If our interest was to study a range of inshore fishery practices, then it seemed advisable to include communities that would allow us to examine a variety of combinations of small-boat and longliner fishermen. Thus, our second criterion was that the communities selected should, as much as possible, reflect a range of differences in the proportion of those who used small, inshore fishing boats and those who used the larger, nearshore longliners. Since few, if any, Newfoundland communities have a predominance of longliner fishermen, in practice this was deemed to mean that we should include both communities in which there were no or very few longliner fishermen, and communities in which there was a significant number of longliner fishermen. The final community criterion had to do with the size of the community. It was anticipated that large inshore fishing communities might display different characteristics of work and community organization than small ones. Consequently, it was deemed important that the communities selected should exhibit a range in both total community size and number of inshore fishermen residing in them.

In addition to these community criteria were criteria related to the choice of fishermen to be studied in each community. Prior to our study, much of the attention in both government reports and the local press focused on the issues of licensing and employment. The fundamental issue here related to the distinction between part-time and full-time fishermen. Thus, the primary prior criterion with regard to the choice of fishermen to be studied rested on the extent to which they were employed in the fishery. It was deemed important that the fishermen selected for interviewing should include a mix of those who worked part time in the fishery and those whose income was derived from full-time employment in it. In choosing this a priori criterion, it was assumed that the extent to which a fisherman concentrated on the fishery rather than practising occupational pluralism would be an important variable on which to base any examination of the impact of state regulation. However, as the interviewing progressed, it was discovered that many part-time fishermen were primarily involved in other occupations and had little knowledge of the fishery as an occupation. Consequently, while some part-time fishermen continued to be interviewed, in the communities to be studied later it was decided to focus primarily on full-time fishermen.

Finally, previous research and the writer's prior knowledge suggested that an important distinction among fishermen was whether or not they held specific species licences. While all fishermen with a part-time or full-time general fishing licence may catch cod, specific species licences give their holders the right to fish for protected species such as salmon, lobster, crab, and herring. Those species tend to be of high economic value, and the fishermen who hold licences to catch them usually receive a much higher level of income than those who do not. Any analysis of the fishery and of the role of state regulation in it should, therefore, include a consideration of the way in which such species licences are awarded and of their impact on work and community life. Thus, it was decided that the fishermen selected for interviewing should include a sample of those who hold species licences and those who do not. Furthermore, as both part-time and full-time fishermen are eligible to hold species licences, where possible the sample of fishermen to be interviewed should include species licence holders from both the full-time and the part-time categories.

Putting the Criteria into Practice

It was still not an easy task to select actual communities and fishermen for inclusion in the study. Indeed, the aforementioned criteria made the task more difficult. To meet them required a detailed knowledge of

Newfoundland communities and their fishing activities. The author might simply have selected communities that his own experience and knowledge suggested were appropriate. However, such a process is fraught with the danger that such 'knowledge' may be inaccurate, or that there are communities, unknown to the author, that would better fit the criteria.

Furthermore, the requirement that fishermen be selected with respect to the licence they held required a knowledge of the licensing structure in a wide range of Newfoundland communities. This the author did not have. To be sure, communities might first be selected and fishermen interviewed within them until a quota of fishermen in various licensing categories was reached. However, the potential for bias involved suggested that this strategy should be a last resort. A better sampling strategy would involve getting a list of all fishery licence holders, by community.

Obtaining a list of licence holders proved difficult. As the subjects of study were fishermen, the most obvious place to go for information was the Newfoundland Fishermen, Food and Allied Workers Union (NFFAWU). Thought the union officials contacted were able to provide detailed information and suggestions concerning which communities might be appropriate to study, a union policy prevented them from releasing any information identifying the licensing structure of communities or the names of the licence holders.[4] A similar request to a senior official in the Newfoundland Department of Fisheries elicited the response that the department had no list of fishermen showing their licensing status. He indicated that the department had been refused access to that information by the Canada Department of Fisheries. Given this, he felt certain that the Canada Department of Fisheries would also be unwilling to allow us such information.

But the Newfoundland regional director of the Canada Department of Fisheries and Oceans proved quite willing to cooperate. He indicated that lists of licence holders by community were available on computer, and that while it would be too massive an undertaking to provide a complete list for Newfoundland, he was willing to provide lists for a relatively limited number of communities. The only stipulation was that such lists should not be made available to anyone else, for any purpose. He also allowed the author to consult with his officials concerning which communities best fit the selection criteria, and, with that information, the author requested and received licensing information for approximately twenty communities. The lists showed the names and addresses of all licence holders in each of the communities for the preceding calendar year. In addition, the lists identified whether a fisherman held a full-time or a part-time licence, and identified any species licences held by the fisherman.

Using these lists, and taking into account the criteria outlined earlier, we eventually chose for study a sample of five communities. It should be emphasized, however, that their selection did not involve a one-to-one correspondence with the criteria for community selection outlined above. Cross-referenced, the three criteria previously identified produced a grid of approximately a dozen different cells. Time and money did not permit us to examine even one community representative of each cell. Instead of attempting to meet the community criteria as strictly defined, we decided to apply them in a more general sense. Thus, we attempted to choose a small number of communities that represented a range of different sizes, within which could also be found differences in the ratio of part-time to full-time licence holders as well as a variety of protected-species licence arrangements.

The lists that we obtained did enable us to divide the fishermen in each of the five communities selected into the following four categories, reflecting the criteria outlined above:

1 full-time licence holders with no species licences;
2 full-time licence holders with additional species licence(s);
3 part-time licence holders with no species licences;
4 part-time licence holders with additional species licence(s).

We were particularly interested in the manner in which species licences were allocated and the implications that such licences had for the organization and practice of fishing activity. Given the small numbers involved, a simple random-sampling strategy might omit or underrepresent those who possessed such licences. Thus, it was decided that a wiser strategy would involve a sampling design that overrepresented full-time fishermen and fishermen who held additional species licences. In each community a random sample was drawn from each of the above categories, with a proportionally larger sample being drawn from the category of full-time licence holders than from that of part-time licence holders. Thus, within each of the four licensing categories, all fishermen had an equal chance of being selected, without any prior bias being at work. However, given our proportional representation sample, those who were in the full-time- and species-licence-holder categories had a greater chance of being selected than those in the part-time- and non–species-licence-holder categories.

Even though this sampling strategy does not permit generalization to all fishermen, we believe that it does allow for an in-depth look at the categories of fishermen who were of greatest interest to this study. Furthermore, it has an advantage over most other studies of fishermen and fishing communities. Most such studies employ a 'snowball' sample, which does not rely on any specific sampling design. Such a method runs the very high risk of focusing on one particular category of fishermen to the exclusion of possible others. The researcher may, for example, interview a group of friends, co-workers, or neighbours whose views may differ substantially from those of other fishermen in the community. Because the latter were not selected, however, the researcher may remain unaware of their alternative perspectives. In contrast, the sampling method used here allows us to be reasonably confident that the fishermen we interviewed were neither drawn from any one clique or friendship group in any community, nor selected using any other overt or covert criteria that might lead to bias. We believe that the sampling technique used here captures a full range of attitudes and opinions within each of the four sampling categories.

The Communities Chosen for Study

Although the five communities chosen for study will be described in detail in subsequent chapters, it is useful to provide a brief description of them here, specifying their locations and the manner in which they reflect the criteria outlined above.

Three of the five communities chosen are located along the east side of Bonavista Bay, on Newfoundland's northeast coast. This area is about 320 kilometres from St John's, and is hence sufficiently removed from the St John's area to have a distinct identity and a largely separate economy, based on the inshore fishery. The area has long been associated with inshore fishing, and, with the exception of some logging and work in local fish plants, has virtually no other source of employment than the fishery.

The three communities chosen were Charleston (1986 population: 242), King's Cove (1986 population: 255), and Bonavista (1986 population: 4605). The distance by road from Charleston to King's Cove is only about forty kilometres, and from King's Cove to Bonavista, about thirty kilometres. Despite this close proximity, the three communities are quite different. Indeed, one of the reasons for choosing them was to demonstrate that communities located quite close to one another can have vastly different work and communal organizations. Charleston is located at the 'bottom,' or the southern inner end, of Bonavista Bay. As large schools of fish rarely come so far 'up' the bay, the location is not particularly desirable for fishing. However, Charleston is closer than other communities in the region to major forest areas, and many residents work as loggers during part of the year. In addition, and somewhat surprisingly, Charleston is one of a relatively small group of communities in Newfoundland that have a fish-processing plant. Charleston was selected for study because of its small size and as an example of a community where a relatively large proportion of fishermen hold part-time licences, practising a combination of fishing and 'woods work' that has traditionally characterized Newfoundland fishermen. Finally, the presence of a fish-processing plant opened up the possibility of examining the impact of such a facility on the occupational and social structure of a small community.

King's Cove is not much larger than Charleston, but it is significantly older, with a proud history as one of Newfoundland's earliest settlements. The large old houses in the community attest to a more prosperous era. Today, King's Cove remains the base for some small-boat fishermen. Most fishermen interested in operating larger boats have moved to the inshore fishing port of Bonavista, only a few kilometres away. Like Charleston, King's Cove was chosen primarily as an example of a small traditional fishing village, but one in which the traditional small-boat inshore fishery is the predominant economic activity.

Bonavista is located near the tip of the headland separating Bonavista and Trinity bays. In addition to being near the very rich fishing grounds of both bays, it is relatively close to large fishing banks some miles offshore. It also has a large natural harbour. The early history of Bonavista and the early history of Newfoundland are virtually synonymous. Indeed, Newfoundland legend has it that the 'discoverer' of Newfoundland, John Cabot, actually landed at Bonavista in 1497, and named the community in his native Italian. Today Bonavista has grown to be one of the largest communities in Newfoundland, and it is the only one of such size to support itself solely on the basis of the inshore fishery. Indeed, it has more licensed fishermen than any other community in Newfoundland. Many of them are small-boat fishermen, but there is also a large longliner fleet prosecuting a mixed-species fishery. In addition, Bonavista is one of a half-dozen communities in Newfoundland that have a longliner fleet licensed to prosecute the crab fishery. The community has both a large regular fish-processing plant and a crab meat processing plant. Bonavista was chosen for study because it offers an opportunity to examine virtually every aspect of a diversified inshore fishery. In addition, it had never before been studied by sociologists and anthropologists. The numerous sociological and anthropological studies of Newfoundland fishing communities that have appeared since the 1960s have focused on smaller fishing communities and on a couple of regional commercial centres. It seemed time that a major fishing centre such as Bonavista received some systematic attention from social scientists. For our purposes, a study of Bonavista opened up the possibility of obtaining new insights into all aspects of both the small-boat and longliner fishery in a large community.

The other two places chosen for examination are also long-established fishing communities. Grates Cove (1986 population: 275) is only slightly larger than King's Cove, and has a history that goes back just as long. Like Bonavista and King's Cove, it owes its origins and continued existence to its favourable location at the very tip of a large peninsula, close to the fishing grounds. However, the most distinctive feature of Grates Cove is its topography. The cove itself is only a slight indentation in a range of towering cliffs that drop two hundred metres to the sea, and the community sits at the top of the cliffs. While the setting is largely inhospitable, the community has long enjoyed a reputation as a place where the fishing is excellent and where the fishermen work hard and are generally successful. Because of the lack of a harbour, it is impossible to operate longliners from the community, but some Grates Cove residents do operate longliners out of nearby Old Perlican. The community has a reputation for being more heavily dependent on 'traps' than many other communities. (As we shall see later, the type of gear used by fishermen proved to be an important factor in our analysis.)

Fermeuse (1986 population: 546) is located on the 'Southern Shore,' a section of coastline running due south from St John's. It is situated roughly one hundred kilometres south of St John's, on an inlet that it shares with neighbouring Port Kirwin (1986 population: 142). It also shares offshore fishing grounds with Renews, situated in the next harbour to the south. Most of the fishermen in the area use the more traditional and smaller inshore fishing boats rather than longliners, but engage primarily in trawling and gill-net fishing. Though some trapping is also practised, it is not the mainstay of the community's fishing activity. In this, and as a mid-sized fishing village, Fermeuse provides a useful contrast to similarly sized Grates Cove, where trapping is the mainstay of fishing activity. It was for these reasons that we included Fermeuse in our study. However, during the time of our study, Fermeuse fishermen were making headlines in Newfoundland newspapers with their complaints about conflicts over fishing grounds and property rights with fishermen from nearby communities, and with their attempts to get the local fish plant to accept fish only from those with full-time (as opposed to part-time) licences. Fermeuse thus offered us an ideal opportunity to get first-hand information on problems involved with licensing regulations and gear conflict.

We have already outlined the underlying dimensions that concern our analyses of these communities. But these factors did not simply exist 'out there,' waiting to be investigated; they emerged from the responses of real fishermen describing their real-life situations. Although the responses naturally depended to some degree on the questions that were asked, our interviews covered a wide range of topics, and our respondents were clearly more concerned with some of them than with others.[5] As we analysed our data, seven topics emerged distinctly as having the greatest relevance for our respondents. While certain of these factors have a higher salience in some communities than others, it is clear that any attempt to explain the nature of fishing activity in these communities must consider all of them. Thus, the seven topics that emerged provided a check-list of dimensions to be discussed in each of our community studies, as follows:

1 the way in which respondents define their occupation and themselves;
2 the extent to which there are community regulations aimed at controlling the allocation of nets and avoiding conflict among fishermen who use different types of gear. This analysis also includes a consideration of the relative frequency of such conflicts in the absence and in the presence of regulations;
3 the impact of all forms of state regulation on the fishery, including such indirect regulatory structures as the unemployment insurance program;
4 the question of whether there is an inherent conflict between holders of full-time and part-time licences and between those who hold protected-species licences and those who do not;
5 the strategies or 'calculus' involved in making a living from the fishery in these communities in the presence of state regulations;
6 the impact of other external agencies, such as the union, fish

plants, and fish buyers, on the work-related activities of the inshore fishermen we encountered;

7 the way in which the residents of each community assess their
quality of life and their community's future prospects.

A Note about Levels of Analysis

A work of this kind involves several issues related to the level of abstraction in the analysis that it contains. In many works these issues
are left implicit, but in a work in which the level of analysis is reasonably complex, as in the case here, there is some benefit in making
more explicit the orientation that is being employed. Three issues are
considered in this regard: (1) the context of explanation involved; (2) the
extent to which the analysis employs a structural or a voluntaristic
orientation; and (3) the level of abstraction at which the analysis takes
place.

The 'context of explanation' can be defined in reference to the communities described in this book. It is possible to understand the five
communities discussed here 'in their own terms.' For example, the
nature of work and community in rural Newfoundland can be understood
and explained only in terms of what is occurring in the communities
themselves. From this perspective, conflicts over the location of gear
are explained in terms of the traditional rights of community members
and in terms of their own perceptions and values. However, such a
perspective has limitations. As we have already noted, the transformation
of work and community in rural Newfoundland is not simply a product
of occurrences within the local communities, but of forces and structures
that extend well beyond them. This problem can be remedied, in part,
by providing contextual information about the prior development of the
local situation and by examining the implications of outside events for
local developments. These tasks are undertaken in chapters 2 and 3.
The problem may also be addressed by examining the implications of
our discoveries in the context of a range of conceptual perspectives, an
approach we have pursued in chapter 4. Specifically, we have used the
conceptual perspectives of property-rights analysis as a prism through
which to observe many of our findings.

The second issue pertaining to the analysis found in this work involves
the manner in which we deal with the eternal tension between structure
and agency (Giddens 1979, 1984). Put simply, do we assume that the
events we observe and describe are determined by social structural

forces largely beyond the control of individuals, or that they are the consequence of the actions of the individuals themselves? Given our earlier discussion of the extent to which state policy affects individual action, along with our statements about the limitations of attempting to explain community change in terms of individual actions and values, it might be expected that this work would adopt a decidedly structural position. That, however, is not the case. While recognizing the enormous power of social structural forces to influence individual action, this work also assumes a voluntaristic perspective that sees individuals as capable of changing the conditions in which they find themselves. In other words, the fishermen who were interviewed for this study and whose words are to be found throughout the following pages are not treated here as mechanistically responding to the requirements of a mode of production beyond their own control (Giddens 1979, 71). Rather, they are seen as individuals engaged in the process of trying to make sense of and manipulate the world in which they find themselves. Their behaviour is treated not simply as the product of their structural position in the society, but as a result of decisions they have made in the context of the structural situation in which they find themselves. Furthermore, our aim has been to examine the role of the state not from the perspective of the dominant classes that might influence its actions but from that of the underclass of fishermen who spend many of their days working within and, often, around regulations imposed by the state. As a result, this book is written neither to demonstrate the power of social structure nor to suggest that individuals construct their own social world, but rather to show how socially situated actors confront their social world and deal with it, often in ways that have unintended consequences both for themselves and for those who are ostensibly in positions of power over them. It is our belief that the task of sociologists is not to focus exclusively on either structure or action, but to achieve an understanding of the dialectical relationship between them.

The final issue concerns the level of abstraction at which the analysis takes place. It is related to the first two dimensions of explanation in that it involves a judgment about how closely one should stick to one's empirical data. But it also involves the question of the level of theoretical sophistication that should be incorporated into the analysis. Although this issue may be approached in numerous ways, one should not lose sight of the fact that explanation always takes place for a purpose (Nettler 1970). This purpose may be as general as attempting to achieve understanding or as specific as attempting to empirically verify a theory. The 'purpose at hand' in this instance is, first and foremost, to explain the way in which life and labour in rural Newfoundland fishing villages are changing. While we certainly employ theory in our analysis, our aim is to use it to help understand a way of life. Consequently, the conceptual analysis found in this work is generally grounded in our empirical data. Theoretical concepts are used when they can help us understand what is taking place in the communities under investigation, or when they can shed more light on the relationship between structure and action as it is found among those who work and live in rural Newfoundland fishing villages.

Conclusion

The preceding pages have discussed the general research context of this work, providing a description of the issues investigated and the research design employed in the community studies. Chapter 2 focuses on the historical context of our analysis. That discussion is followed by a more detailed examination of the fishery policy context in chapter 3 and an in-depth investigation of the conceptual context in chapter 4.

2 The Historical Context

The received image of the poor fisherman is to be stood on its head – he is a fisherman because he is poor, not the other way around.

W.C. MacKenzie,
'Rational Fishery Management in a Depressed Region' (1979, 816)

In one of his books on the Newfoundland fishery, Peter Sinclair (1985, 31) states:

Although immediate political and economic factors (such as fisheries management regulations and recent fish prices) are often decisive in generating and moulding change, the accumulated experience and consequences of past practices set the stage upon which contemporary forces play: the past limits what is possible in the present and influences what is perceived to be possible.

Although it has become a commonplace to observe that historical background gives context to current events, the observation is particularly apt in the case of the Newfoundland fishery. The relevant historical factors include, in addition to the historical development of the fishery itself, aspects of general Newfoundland history relating, in particular, to the transformation of work and economy in the province. This chapter's brief examination of the relevant historical factors provides the context for our subsequent analysis.

The General Historical Context

Social change takes place everywhere, but the pace of change and the impetus for it to occur vary tremendously. Some places change gradually, others rapidly; some change as a result of their own actions, while others have change thrust upon them by external forces largely beyond their control.

Throughout most of Newfoundland's history the pace of change has generally been extremely slow and gradual. The island was first settled by Europeans in the early 1600s, and for the most of the succeeding 350 years, little change occurred. St John's, the capital city, remained a small commercial centre, while most of the population continued to reside in small and often isolated 'outport' communities, carrying out their daily activities in much the same way as their forefathers had. In most such communities, fishing was the basis of daily life. But because fishing rarely provided much more than a bare subsistence, most families also engaged in other economic activities, often subsistence activities such as maintaining a garden and a woodlot for personal use. In some communities on the northeast coast, fishermen also participated in the seal hunt.

Most outport fishermen were bound to the outside world through two independent sets of socio-economic relationships. One was a set of hierarchical dependency relationships (Gunder Frank 1969, 21–94; Matthews 1983) that operated as follows: firms in England provided supplies on credit to Newfoundland firms based primarily in St John's, which in turn provided them on credit to local village merchants. Those merchants provided goods on credit to local fishermen and in return received dried fish at the end of the season. As the merchants, at each level, set the prices of both the fish and the goods they exchanged for them, few fishermen ever managed to free themselves of debt. The second set of socio-economic relationships arose out of a form of patron–client organization (Paine 1971; Long 1977, 43–5, 119–26) in which local village merchants were the patrons, negotiating with the outside world on behalf of their clients, the village fishermen and their families.

Three major developments in this century have been largely responsible for altering the basic socio-economic character of Newfoundland: (1) the development of logging and paper making, (2) the Second World War, and (3) Confederation with Canada.

Logging became a major activity in the period 1890–1920, with several major lumber mills opening throughout the northeast and western areas of the province (Thoms 1967b). The logging operations provided a major source of supplementary income for fishermen, who worked as loggers in the winter and spring. These operations also provided the basis of a workforce for two major paper mills, the first started at Grand Falls in central Newfoundland in 1909 (Tucker 1975) and the second at Corner Brook on Newfoundland's southwest in 1925 (Smallwood 1975). These mills were significant in two ways: first, as already noted, they provided supplementary income from logging for hundreds (if not thousands) of Newfoundland fishermen; second, they led to the development of two urban centres outside St John's. Winter logging offered fishermen their first break from a complete dependence on the fishery for their livelihood. Furthermore, the cash income from logging gave them their first opportunity to circumvent the barter system dictated by the village fish merchant.

The Second World War had an even greater impact on Newfoundland's economy and society. For centuries an isolated backwater in the flow of world trade, Newfoundland suddenly became a strategic location in the war effort. Its harbours (notably St John's) provided safe supply and refuelling points for shipping convoys carrying North American goods for the Allied war effort. The result was an influx of cash into the local economy and a supply of new jobs, both on shore and aboard ships. But Newfoundland's strategic location[1] also made it a potential 'first line of defence' in any attempt by Germany to invade North America. The most significant wartime event, as far as the social and economic structure of Newfoundland was concerned, was the construction of four large U.S. air and naval bases on the island. Because most American personnel were wanted for the war effort, the task of building the bases and providing the staff needed to run them went to Newfoundlanders. The American forces provided training and previously unheard-of salaries to their local employees. Furthermore, many Newfoundland women married U.S. servicemen stationed on the new bases. Thus, by war's end, thousands of Newfoundland men had marketable skills, and thousands of Newfoundland women found themselves transferred, with their new husbands, back to the United States.

In a brief four-year period rural Newfoundland had been transformed from, essentially, a barter economy and an isolated society into a cash-based economy and a more outward-looking society. Few of those with newly acquired technical skills were willing to return to their previous way of life in remote and isolated fishing outports. Some migrated to Canada and the United States to find work. Others remained on the bases, which were not phased out until the 1960s, or found employment in St John's and other service centres.

The third major factor that had a dramatic role in shaping Newfoundland society was the decision by its people, in 1949, to enter into Confederation with Canada. Newfoundland's history as a separate political entity dates back to 1855, when it was granted the right by the

British government to establish responsible local government. In most respects, Newfoundland was essentially self-governing until 1933, when its economy went into complete collapse, brought on by the Great Depression and a declining world market for Newfoundland's major staple commodity, salt fish. However, mismanagement and corruption by local governments also contributed to the collapse. Indeed, a British royal commission established to determine what to do about the situation recommended that the people of Newfoundland be given 'a relief from government' on the grounds that the society could no longer afford the corrupt activities of its local political and economic élite (Great Britain 1933). The commission recommended the appointment of a 'Commission of Government' – a body that would govern the colony until economic and political order could be re-established. Thus, from 1933 to 1949 Newfoundland was ruled by a six-person commission government appointed by the British Crown.

By the mid-1940s, economic order had been re-established in Newfoundland,[2] and the commission government deemed it time to return the island to democratic rule. As a step in this process it held a 'National Convention' in St John's, at which newly elected representatives debated the alternatives of becoming self-governing, joining Canada, or remaining under the commission government. Two public referenda resulted from these meetings. The first defeated the option of a return to self-government, and the second upheld Confederation with Canada by a small margin.

The decision to join Confederation can be interpreted as a response to the charismatic leadership of Joseph R. Smallwood, who had championed the cause of Confederation and who became the new province's first elected premier, a post he held for more than two decades. On a more social-structural level, the decision can also be attributed to the changes that Newfoundland society had undergone since 1933 (Matthews 1979). Confederation with Canada now offered the immediate additional advantages of a modern welfare state, complete with old age pensions and family allowances.[3]

The period from 1949 to the mid-1980s saw a massive transformation of Newfoundland society. All of the infrastructural amenities of a modern society were introduced, bringing the province's social and personal services up to the level that prevailed in the rest of Canada. During this period, St John's became a major service centre.[4] However, the influx of massive transfer payments from the federal government altered Newfoundland's society (and economy) in rather unusual ways.

Essentially, Newfoundland has developed into a service economy without the industrial and economic base that is necessary to support it. An indication of this problem is apparent in the labour-force statistics. For example, the number of people employed in service, or tertiary, sector activities under the categories 'Community, Business and Personal Service' and 'Public Administration' has increased significantly, from 21,247 in 1951, when workers in these sectors constituted 19.9 per cent of the labour force, to 78,310 in 1981, when they constituted 35.3 per cent of the labour force. By contrast, the 'Trade' and 'Manufacturing' sectors grew only marginally during the same period, rising from 13.9 per cent to 16.1 per cent and from 13.0 per cent to 16.3 per cent of the labour force, respectively.[5] Furthermore, employment in the primary activities of farming and logging has also dropped severely. Employment in agriculture has declined from 3.3 per cent to 0.6 per cent of the labour force, while in forestry (which was hard hit by the mechanization of the logging industry) it has declined from 9.9 per cent to only 1.8 per cent of the labour force.

It is difficult to get accurate and reliable statistics on the number of people employed in fishing during this period.[6] None the less, it is clear that there has been considerable fluctuation. Using data on the number of licences issued, McCorquodale (1983, 154) indicates that in 1980 there were 35,080 licensed fishermen in Newfoundland, up significantly from a low of 12,792 in 1974. According to her, between 1974 and 1979 the number of inshore fishermen in Newfoundland increased by 170 per cent. House (1986, 23) has provided the following explanation for this growth, together with an update on changes that have occurred since 1979:

After a prolonged decline in the number of fishermen in Newfoundland, the combination of restricted alternative opportunities and high expectations from the 200-mile limit saw a rapid increase until the boom year of 1979. Since then, the numbers have declined again ... The decline since 1981 has been accounted for entirely by part-timers. Their numbers fell by 23.7 per cent, from 15,212 in 1981 to 11,611 in 1985. The numbers of full-timers has been steady at just over 13,000. A DFO [Department of Fisheries and Oceans] study of *active* licensed fishermen estimates that numbers so defined fell by 18 per cent, from 22,870 in 1981 to 18,658 in 1984.

Even with these declines, it is clear that fishing still constitutes a major source of both full- and part-time work in Newfoundland, particularly in the rural regions.

The Development of the Newfoundland Fishery

In many ways, Newfoundland's history and the history of fishing in Newfoundland are virtually inseparable. The Norse came to Newfoundland somewhere around the year 1000, and there is now abundant evidence that the waters off the island were being used regularly as a fishing-grounds long before its official 'discovery' by Cabot in the late fifteenth century. Certainly, by the early 1500s many ships from England, France, Spain, and Portugal were regularly visiting Newfoundland (Judah 1933, 15–17; Rogers 1911, 25). By the mid-1700s some ten thousand men were coming annually from Devon and Dorset alone to fish off Newfoundland's shores (Matthews 1968, 10), but hardly any chose to settle there. The most popular explanation for this is that settlement in Newfoundland was forbidden under a 1633 British Star Chamber Rule (Perlin 1937, 173), but after this rule was effectively rescinded, in 1677, there was no significant increase in population. The more likely explanation for the lack of settlement is that there was no reason to live in Newfoundland year-round. Quite simply, Newfoundland was 'nothing but a fishery' (Matthews 1968, 10–15), and it lacked any source of winter employment. Permanent settlement dates from the early nineteenth century, when changing European clothing fashions led to a demand for sealskins in Europe, thus providing a basis for springtime employment in Newfoundland's seal hunt. During the same period, Britain's involvement in the Napoleonic Wars, the American Revolution, and battles in India required all its merchant navy as supply vessels, and it was willing to encourage settlement in Newfoundland in order to ensure a continued supply of fish (Matthews 1970, 23–27). This brief history is important because Newfoundland outports, in one respect at least, are little different today from the way they were in the early nineteenth century: For the most part, they lack any basis on which to sustain winter employment. Consequently, they are always in danger of being judged 'non-viable' by policy makers.

The original settlers in Newfoundland tended to locate in coves close to fishing grounds. As the population grew (largely through natural increases), the people tended to migrate along the shoreline in search of new fishing grounds. Generally speaking, the best fishing grounds are to be found around the offshore islands and off the most distant headlands. The result was a widely dispersed population living in hundreds of small communities. As long as the primary mode of travel was by sea and there were few public services such as roads, electricity, schools, and hospitals, this did not present much of a problem. But with the massive increase in infrastructural services that followed Confederation, many of Newfoundland's widely dispersed communities suddenly came to be considered 'isolated.' It was this that led to the government-sponsored community-resettlement program of the 1960s and the 1970s (Iverson and Matthews 1968, Matthews 1976) that aimed to consolidate the population in 74 selected growth centres. Although the resettlement program succeeded in phasing out well over a hundred communities,[7] 625 communities, each with a population of less than 5000, remained in 1976 (Kirby Task Force 1982, 70). At that time, the total population of Newfoundland's fishing communities was estimated to be about 280,000 (Kirby Task Force 1982, 71), which represented significantly more than half the population of the province.

As for the fishery itself, though Newfoundland waters have generally had an abundant variety of fish, until recent years there has been little demand for anything but cod, lobster, and salmon. Of these, the cod fishery has been the primary fishery throughout most of the summer fishing season.[8] The traditional methods for catching cod were the handline, trawl, and net, but the invention of the cod trap in the late nineteenth century and the invention of gasoline engines in the 1920s provided a more efficient alternative technology.[9]

Historically, the small-boat inshore cod fishery was organized around the extended family. Its social organization consisted of two different production units. The catching or harvesting of fish was carried out by the small-boat crew, which usually comprised two to four males from the same extended family, for example, father and sons, or brothers (Nemec 1972). It was their job to 'hunt' and catch the fish, and to gut, split, wash, and salt it on the 'stagehead' once it had been brought to shore. The rest of the processing was carried out by a processing unit consisting of the remainder of the extended family – the women and children (Porter 1985, 115–16). The women and children were responsible for spreading the fish to dry in the sun in good weather, and for collecting and storing it at the end of the day, or whenever rain threatened.

While elements of this traditional organizational structure still remain throughout all parts of Newfoundland, the past twenty years have seen major changes in the technological development of both the harvesting and the processing of fish. These, in turn, have brought about a considerable transformation of the social organization of fishing. For the purpose of our analysis in this book, three aspects of this transformation are of particular relevance: (1) the transformation of harvesting through the introduction of longliners and the development of a 'nearshore' fishery; (2) the transformation of processing through the development of fresh-fish-processing facilities; and (3) the transformation of access through licensing and limited-access regulations. The first two are the subject of the remainder of this chapter. The issues of licensing and access are so complex, however, that they require an extensive discussion. They are the focus of our attention in chapter 3.

The Transformation of Harvesting: The major transformation in the harvesting sector involved the introduction of longliners. The traditional fishery used boats that were generally less than thirty feet in length and could operate within a few kilometres of shore, but were clearly limited to fishing grounds within sight of land. Such boats could not operate in the bad weather of early spring and fall; they could not safely take advantage of fish stocks that were known to exist in shoal waters several kilometres offshore; they could not carry the large amounts of fishing gear that would make more distant fishing profitable; they could not carry back the large quantities of fish that would also be necessary to make more distant journeys paying propositions; and they were not sufficiently large to legitimize the cost of modern 'fish-finding' gear, such as depth sounders and advanced electronic navigational aids.

As a consequence, beginning in the early 1960s, the Newfoundland government began to encourage the use of larger vessels known as 'longliners,' which were supposedly capable of fishing the 'nearshore.' The nearshore was loosely defined as the area beyond the inshore in which the smaller-boat inshore fishermen could safely operate, but which was still inside the 'offshore' banks fished by the large trawlers and draggers. To encourage the development of this fleet, the Newfoundland government provided a variety of types of assistance, including low-interest loans and loans that reduced in value with each year the boat was in operation. As a result of such incentives, approximately one-fifth of all Newfoundland inshore and nearshore fishermen now fish from longliners.

The introduction of new technology, in the form of longliners, brought with it a transformation in the nature of work in the Newfoundland inshore fishery. Although the longliners were capable of fishing well beyond the limited area to which the inshore boats were confined, there was nothing forcing them to do so, and some longliner operators indeed chose to fish the inshore waters in competition with the small-boat fishermen. Their size and the sophistication of their fish-finding gear naturally made them far more efficient than the inshore boats. In addition, the longliners tended to specialize in the use of gill nets, and one boat could tend more than 150 nets. A large number of nets placed in inshore waters near good cod-trap 'berths' could capture all the fish on which the inshore trap fishermen depended.[10]

The introduction of longliners also transformed the social-class relations that had previously existed in the inshore fishery. The inshore boats were often owned in common, or a single crew might operate two or more boats, of different sizes, shapes, and capabilities,[11] each owned by a different member of the crew but used in common. In most cases the income from the catch was divided equally, with 'the boat' also being accorded a share to offset costs of upkeep, fuel, and gear. If a crew included 'boys' who were essentially apprentices, or 'sharemen,' who had no ownership claim to the boat, they were usually contracted for a portion of a share. Thus, from a social-class perspective, although the crew members were not all equal in terms of investment and rewards, they were equal in terms of their burden of risk. Furthermore, no member of the crew could in any real sense be construed a capitalist, since no member accorded to himself the surplus value of the labour of the other members of the crew.[12]

Although the terminology used in reference to longliners is much the same, the patterns of their ownership are quite different. A longliner is usually owned by an individual 'skipper,' who must bear the responsibility for the loans and mortgages that are outstanding on the boat. His crew members are still called 'sharemen,' and the boat is still accorded a share, but the greater investment in the boat usually means that the 'boat share' rises to approximately 40 per cent of gross receipts. When fishing is good and prices are high, the skipper retains any surplus from the boat share as his own additional profit. Fairley (1985, 40) has argued that longliners should therefore be considered capitalist enterprises. He uses Department of Fisheries and Oceans data from 1960 to show that, on the five longliners he studied, skippers earned a profit of 13.5 per cent on net capital invested. He also observes that 'The capitalistic character of the nearshore phenomenon is more clearly indicated when it is examined as a social process of production ... Thus of the total fifteen persons employed, seven brought both labour power and means of production into the production process, while eight brought only their labour power' (39). Fairley argues that longliners remain capitalist enterprises even when the 'boat share' fails to make a profit for the boat owner: 'They should be seen as enterprises that were capitalistic in ownership structure and mode of surplus appro-priation, but ones that failed, as many capitalist enterprises periodically do, to provide an adequate return to their owners because of catch-levels achieved, species-mix pursued and caught, and/or prevailing fish prices' (41).

While Fairley's argument may hold true in terms of Marxist definitions of capitalist relations, a more culturally sensitive interpretation can be found in Peter Sinclair's work on shrimp fishermen who fish from longliners (1985). Sinclair notes that, even though longliner enterprises have many of the characteristics of capitalist enterprises, since 'the skipper is an owner-operator of a small business, directly comparable in this respect to the farmer' (86), they are also different in important ways. For example, employees are often chosen on the basis of kinship (97); employees still receive a share of the catch (98);[13] and skippers cannot be said to exploit labour in the sense of expropriating from the sharemen the value of what they produce, 'for their earnings are much in excess of any other available work for semi-skilled labourers' (99). Although Sinclair's position seems to imply, questionably, that relatively high wages preclude the possibility of exploitation, his aim is to argue that the Newfoundland inshore fishery, rather than being seen as a purely capitalist system, is more appropriately viewed as a prime example of domestic commodity production. The longliner operation may best be described as dependent petty capitalism, in which most of the boat owners 'have personal biographies rooted in domestic commodity production and have retained aspects of domestic commodity production that are congruent with petty capitalism' (142).

Although such changes in the social relationship of ownership and employment in the Newfoundland inshore fishery have been discussed by Fairley and others in the general theoretical literature, there has been very little research that has attempted to examine the way community life and the nature of work have been transformed as a result. Sinclair's is virtually the only major study that has attempted to do so. However, despite its sensitivity to the local issues, it is limited by its focus on only one community (Port aux Choix) and one very distinctive type of fishery (the shrimp fishery). Given the unique and specialized character of that fishery, it is not clear that we can generalize from Sinclair's findings to other communities and other types of fisheries.

The community studies that follow are, in part, an attempt to examine how the nature of the inshore fishery and the nature of the communities in which it is based have been altered by changes such as those just described. More specifically, we examine how relationships between inshore and nearshore fishermen using different 'technologies of production'(Matthews 1983, 194–8) are handled in a community context. We also attempt to depict the changing nature of class relationships among fishermen who, at one level, are all equal members of the same union of workers, but, at another level, have clear differences in terms of class interests. Since the latter may distance them from one another, it may be a myth to call the local fishing outport a 'community,' given the implications of common interest and unity that are normally associated with that term.

The Transformation of Processing: Not all the changes that have taken place in the Newfoundland inshore fishery over the past twenty-five years have been confined to the harvesting sector. The processing sector has also undergone a massive transformation. The key factor here has been the virtual abandonment of the practice of salting fish, and the development of an extensive network of fish processing and freezing plants in its place. The number of fresh fish processing facilities in Newfoundland increased gradually over the past thirty years, and skyrocketed during the 1980s. In the period 1977–81, 108 new plants were constructed in Newfoundland, bringing the total to 255 (Kirby Task Force 1982, 31). Since then, a combination of failing markets and overexpansion has brought about a number of plant closures and a reorganization of the ownership structure of the inshore fishery.[14] Nevertheless, in 1985, 202 fresh fish processing plants remained along with an additional 18 mechanized salt-fish drying operations (calculated from data in House Royal Commission 1986, 130). Most of the fish caught by Newfoundland fishermen is now processed and fresh frozen. Furthermore, the processing facilities are able to deal with virtually all types of fish, with the result that there is now a market for species that fishermen had previously ignored. Some of these, such as shrimp and crab, require special equipment and are processed in separate plants. Others, such as sole, flounder, haddock, halibut, and redfish, are processed in the same plants as the ones used for cod.

From a social point of view, the fresh fish processing facilities have had their greatest impact on the labour-force structure in rural Newfoundland. There were 23,282 fish-plant jobs in the province in 1980, and 25,021 in 1984 (House Royal Commission 1986, 124–5). However, since job turnover in such plants is high, many more individuals than these numbers indicate actually benefited from fish-plant employment. There were nearly twice as many fish-plant jobs as there were full-time fishermen in Newfoundland in those years, and almost as many fish-plant jobs as the total number of full- and part-time fishermen combined (based on data in House Royal Commission 1986, 195). Considered from a slightly different perspective, 20–25 per cent of fishermen's households in Newfoundland had one or more members working in fish plants in 1981 (Kirby Task Force 1982, 67).

Some of the existing analyses of the transformation of the processing industry have emphasized its negative consequences for the fishermen. For example, it is argued that the conditions of work for fishermen as independent commodity producers have deteriorated, in that the fishermen no longer have the freedom to control their own labour, but are now bound by the buying practices of the fish plants and the quality and size regulations they impose (Connelly and MacDonald 1983, 54; Fairley 1985). However, while the fish plants are indeed large capitalistic enterprises with all the attendant labour-relations difficulties, such statements would seem to be colouring the traditional fishery with far too rosy a hue. In many cases, the only freedom that the fisherman had in the traditional fishery was the freedom to starve, and he was often totally dependent on the whims and good will of the village merchant. In fact, the availability of fish-plant work has had a particularly positive impact on fishermen who had previously worked part time as loggers and sealers. The decline of the logging industry as a source of part-time employment during the 1970s severely threatened many such fishermen, in that it left them without the additional income that made their part-time fishing viable. The collapse of sealing in the early 1980s as a result of the activities of animal-rights activists had a similar impact on some northeast-coast fishermen. Part-time work in fish plants has provided a welcome alternative source of income to supplement part-time fishing income. Without it, many part-time fishermen might well have been forced out of the fishery altogether, a development that would have undermined the economic viability and, consequently, the social vitality of the communities in which they lived. Part-time employment in the fish plants has been particularly beneficial to middle-aged and older fishermen who, through illness or the infirmities brought on by long years of hard physical work, no longer have the strength or stamina to engage in the fishery full time, but who have the skills needed to act as cutters and splitters in fish plants.

The greatest impact of the shift to automated fish processing has been on the women of outport Newfoundland, the wives and daughters of the fishermen. Where previously their contribution to the processing of fish brought no direct financial rewards, they can now get immediate cash payments for their labour as trimmers and packers in fish plants. The implications of this changing role of women in rural Newfoundland have received little attention from researchers. The women of Nova Scotia, however, are now the subject of intensive study (Connelly and MacDonald 1983; MacDonald and Connelly 1986a, 1986b).

The Nova Scotia studies emphasize the extent to which women who are employed in fish-processing operations find their work unsatisfactory. The women interviewed in these studies complain about being fixed to the line positions as trimmers and packers, while the men are able to move around in their activities as cutters and splitters (Connelly and MacDonald 1983, 63). The authors also point out that the relatively fixed plant hours of female fish-plant workers do not mesh well with the odd and often erratic work hours of their fishermen husbands, making it difficult for them to look after their children and perform the household duties that their husbands still expect of them (MacDonald and Connelly 1986a, 9). The women surveyed also indicate that fish-plant work does not provide sufficient remuneration to offset the lost income-tax deductions, the extra income tax payments, and the additional costs involved in working (1983, 61). It is likely that Newfoundland women employed in fish plants would have similar complaints; in the Newfoundland context, however, those concerns may be overshadowed for many women by the 'benefits' of such employment. Not only does seasonal fish-plant work bring in much-needed income, it also permits many women to receive unemployment insurance benefits throughout the remainder of the calendar year.

Because of the high unemployment rates and the seasonal nature of much of the available employment in Newfoundland, federal regulations allow workers to qualify for unemployment insurance after only eleven to fourteen weeks of continuous work.[15] This means many female fish-plant workers can turn less than four months' employment into weekly unemployment insurance payments throughout the balance of the year that may, in total, exceed their employment income. Ironically, their fishermen-husbands may not be as lucky. In their case, qualification for unemployment insurance depends on the quantity of fish they bring for sale during the fishing season. If the total catch in a season is poor, they may not qualify for unemployment insurance at all, and are thus, in effect, doubly penalized. Furthermore, unemployment insurance of fishermen is granted only for the period from January to May. In years when ice blocks the coastline well into June or when the fish are late in arriving, the fishermen may have to face several weeks without income after their unemployment insurance 'runs out.' In such cases, the weekly employment cheque brought in by their wives may be the family's main source of income. From the women's perspective, however, having their own earned income has perhaps afforded them a new independence and control over their lives that would previously have been impossible.

Conclusion

This chapter has provided the background information about Newfoundland that is necessary to understand recent developments in the East Coast fishery. In particular, it has emphasized the extent to which the social organization of community and work in rural Newfoundland has been transformed by recent changes in the society at large and, more specifically, in the organization of harvesting and processing fish.

In addition to changes in the technology and social organization of the harvesting and processing sectors, the Newfoundland fishery in the past three decades has been subject to major changes in the regulations governing 'rights of access' – that is, rights to the catching and selling of various species of fish. This 'policy context' is examined in chapter 3.

3 Changing Metaphors

Value Reorientation in Canadian Fishery Policy

A world ends when its metaphor has died.

Archibald MacLeish,
'The Metaphor'

Fisheries policies, like all policies, are 'goal-value systems'; that is, they constitute value orientations that dictate the appropriate means for achieving desired goals.[1] Consequently, in order to gain a comprehensive understanding of the nature of fisheries policies for Canada's East Coast, it is important to understand the changing value orientations that underlie such policies. Our treatment of the subject therefore goes beyond a simple chronology of changes in Canadian and Newfoundland fisheries policies to a consideration of the reorientation of values and goals that may have prompted those changes.

The fishery, like many forms of human activity, can be regulated in two ways: through a system of customary and largely local practices or through a system of formally constituted rules, regulations, and laws created by the state. While it is possible to envisage a time, before the rise of the centralized state, when the regulation of local fisheries was subject largely to use-rules developed in and enforced by local fishing communities, that time is long since past. Today, the nature and the form of state regulations set the parameters within which local regulation is possible. But, as we shall see later in the book, the customary, local ways of regulating the fishery have not disappeared. They continue to exist and frequently provide an alternative set of rules that remain 'nested' within the more formal regulations emanating from the state. Some of the surviving customary rules present few, if any, problems, because they deal with aspects of the fishery that are not covered by state regulations. Others, however, conflict directly with the official rules.

The conflicts that arise usually involve a difference in visions of the nature of the fishery. They revolve around different outlooks about the extent to which local fishermen can adequately conserve the fish stocks, and involve different views about the nature of fishery organization and the types of institutions that are required to regulate it. Modern social science has tended to describe times when conflicting views of the world collide as periods of 'paradigm change' (see Kuhn 1970). Unfortunately, the term *paradigm* has, over the years, taken on such a wide variety of meanings as to have been rendered virtually meaningless (Eckberg and Hill 1979). Moreover, as it is generally used, the concept of paradigm fails to capture the idealistic quality of the competing value orientations underlying the different views of appropriate fishery policy. Those competing visions of the nature of the fishery have a certain metaphorical quality: They are not descriptions of reality as much as they are 'ways of seeing' reality.

This chapter, then, is about different visions of how the fishery on Canada's East Coast should be regulated. It documents how a metaphor relating to the supposed 'tragedy of the commons' came to be accepted as the basis of Canadian fishery policy – at least at the federal level. It also demonstrates that there is, to some extent, a conflict of values and visions between the federal government and the Newfoundland provincial government with respect to the regulation of Newfoundland's inshore fishery.

The Changing Value Orientation of Canadian Fishery Policy

Several works have appeared in recent years outlining the development of Canadian East Coast fishery policy (see Barrett 1981; Copes 1980, 1983; Davis and Thiessen 1986; Draper 1981; Hanson and Lamson 1984; House 1986; MacDonald 1984; Martin 1973, 1979; McCorquodale 1983).[2] This recent spate of work on Canadian fishery regulation is to some extent attributable to the considerable involvement of the Canadian government in this area since the early 1960s. Most of the existing works focus on developments since the early 1960s, providing only brief summaries of earlier political activity, usually beginning with the assignation of 'sea coast and inland fisheries' to the federal government under the British North America Act in 1867.

Many of the recent studies see Canadian fishing policy as having passed through a series of periods, stages, or cycles, each with its own orientation. For example, McCorquodale (1983, 156) argues that it is useful to think of fishery policy in terms of cycles that are determined in large part by biological and economic factors, and, on that basis, she divides Canadian fishery policy into three periods: before 1950, 1950–70, and after 1977.[3] MacDonald (1984) is concerned primarily with issues of administration within the Canadian Department of Fisheries and Oceans. He describes the period from 1945 to 1977 as one during which administrative and decision-making structures were put in place, and the period from 1977 to 1981 as dealing largely with issues of extended jurisdiction and the 200-mile limit. Barrett (1981) focuses on the extent to which the state took an active interest in regulation. In contrast to MacDonald and most other observers, he tends to see the period from 1939 to 1973 as characterized by little state involvement in the fishery. Thus, he identifies three phases: 'a strongly regulatory phase between 1930 and 1939, a "laissez-faire" phase lasting from World War II to 1973, and the period of renewed regulations since 1974' (1981, 1). Similarly, in the most comprehensive review of fisheries policy available to date, Drapper (1981, 111) identifies three major periods, as follows: 'The first, Confederation to 1965, was characterized by a biological emphasis. Responses to fisheries problems and pressures were primarily reactive and ad hoc. The second phase, 1965 to 1976, saw the addition of economic and social considerations to the biological basis of management. The recent period, 1977 onward, is associated with implementation of the "200 mile limit."'

All the analyses noted agree that there have been significant changes in the direction and value orientation of Canadian fishery policy. Most emphasize a significant change in the value orientation beginning in the late 1950s or 1960s, when the focus of concern shifted from the purely biological aspects of fish-stock conservation to a broader consideration of the social and economic aspects of the fishery. All the studies document a further change in focus around 1976, after Canada won jurisdiction over a 200-mile limit at the International Law of the Sea deliberations.[4]

The Origins and Assumptions of the Common-Property Perspective on Fishery Regulation

The shift in policy direction during the 1950s and the 1960s appears to have been directly related to the development of the economic theory of common property as applied to the fisheries. Although the works to which we refer above use a variety of criteria to identify various policy periods, they do not, in general, focus on the extent to which the common-property perspective transformed the way regulation of the fishery was perceived. Furthermore, while these studies focus on *outcomes*, in terms of the stages of fishery policy, the present discussion examines the *process* involved in the formation of such policies. We argue that the common-property perspective gave rise to a new way of viewing the fishery and, consequently, regulation of the fishery. Before the common-property perspective gained acceptance, regulation of the fishery was concerned primarily with biological conservation. By contrast, the common-property perspective treated the fishery as an economic and social system that had to be regulated on the basis of considerations other than just the biological.

The economic theory of common-property as applied to the fishery has its origins in two major works. The first is Scott Gordon's analysis (1954). Gordon argued that the fishery was a common-property resource, to which all persons had rights of access.[5] He suggested that, under circumstances of open access, it would not be rational for an individual fisherman to refrain from fishing in order to ensure the conservation of the fish stocks. Should he do so, others would undoubtedly take advantage of his actions and capture his share of the available resource. Gordon concludes that 'there appears ... to be some truth in the conservative dictum that everybody's property is nobody's property. Wealth that is free for all is valued by none because he who is foolhardy enough to wait for its proper time of use will only find that it has been taken by another ... The fish in the sea are valueless to the fisherman, because there is no assurance that they will be there for him tomorrow if they are left behind today' (124). Consequently, the rational action for each individual in a common-property, open-access fishery would seem to be to take as much catch as possible before it is appropriated by others. The only alternative Gordon could envisage involved some form of regulation; the commons must in some way be 'fenced' in order to turn it into private property. As he put it, 'This is why fishermen are not wealthy, despite the fact that the fishery resources of the sea are the richest and most indestructible available to man. By and large, the only fisherman who becomes rich is one who makes a lucky catch or one who participates in a fishery that is put under a form of social control that turns the open resource into property rights' (132).

The other fundamental work in this area is by Garrett Hardin (1968), who supported and extended Gordon's position to all situations in which a form of common property exists. The result of such a system of ownership, he argued, is invariably the depletion of the common resource. In his words, 'Ruin is the destination toward which all men rush, each pursuing his own best interests in a society that believes in the freedom of the commons. Freedom in the commons brings ruin to all' (1244).

As far as Canada's East Coast fishery was concerned, the clear implication of such analysis was that a focus on only the biological aspects of conservation would be insufficient to conserve fish stocks. Economic and social regulation, in addition to biological regulation, was desperately needed.

With the acceptance of common-property theory, economists came to play a major role in developing Canadian fishery policy. Moreover, fishery biologists (who had until that time been the main players in formulating government fishery policy) quickly adopted the common-property perspective and made it the basis of their own management policies.[6] Copes (1980, 132) states that 'serious economic analysis of the common property problem of the fishery started only in the mid 1950s and it took a decade for useful policy prescriptions to work their way into government plans.' House (1986, 2) expanded on this point, arguing that 'while Canadian fisheries policies did not come to fruition until the late 1970s and the early 1980s, the philosophical and theoretical seeds for these polices were sown much earlier by economists such as H. Scott Gordon and Parzival Copes. A set of assumptions and principles emanating from fisheries economics has influenced the thinking of federal and provincial officials and policy-makers and, to a large extent, the general public.'

For those interested in the nature of scientific explanation, or what is sometimes referred to as the sociology of science, the shift in perspective just described represents an interesting case study of the displacement of one discipline's point of view by that of another. While it might seem logical to assume that biologists would have a virtual monopoly on claims of expertise in matters pertaining to the conservation of fish stocks, the impact of the common-property work of Gordon and others was apparently so profound that biologists' claims to authority in this area were being successfully challenged by economists.[7] So pervasive was this process that, as noted above, fishery biologists themselves assumed the 'tragedy of the commons' perspective as the basis of their own approach to resource management. Even more interesting is that this particular economic theory was based essentially on social and psychological variables having to do with the nature of human motivation under conditions of common property. Such issues are usually the preserve of anthropology, sociology, and social psychology, but specialists in these disciplines were unable to capitalize on the new developments the way economists did – that is, they were unable to increase their presence and prominence in the area of fishery policy and regulation.

Gordon's and Hardin's work provided the basis for an outpouring of study and analysis. Numerous works appeared throughout the 1960s and 1970s on the implications of the common-property nature of the fishery.[8] This period of conceptual development can be seen as culminating in two conferences sponsored by the Institute for Marine Studies at the University of Washington (see Mundt 1975, Retting and Ginter 1978), and a symposium published in the *Journal of the Fisheries Research Board of Canada*, which contained articles by leading fisheries economists (see Crutchfield 1979, Fraser 1979, MacKenzie 1979, Moloney and Pearse 1979, Pearse and Wilen 1979, Scott 1979, Wilen 1979).

The symposium papers in the *Journal of the Fisheries Research Board* deserves our careful consideration, for two reasons. First, they provide a series of statements and some level of synthesis of the ideas developed by Canadian fisheries economists over the course of the preceding two decades. In that sense they are a good indication of the 'state of the art' in fisheries economics in the mid-1970s. Second, as we noted earlier, most analysts who have reviewed the development of Canadian fisheries policy agree that there was a change in direction beginning in the mid-1970s. These papers provide an excellent insight into the ideas and values that formed the basis of that shift in the value orientation of Canadian fisheries policy.

A review of this body of literature indicates that five themes formed the basis of the common-property perspective on fisheries policy.

First, there was total and universal acceptance that a common-property, open-access fishery leads to overcrowding and the depletion of the fish stocks. As MacKenzie (1979, 811) succinctly stated, 'it is accepted on theoretical grounds that under conditions of open access the tendency to overcrowding and depression is universal and inexorable.'[9]

Second, there was general agreement that measures were necessary to move excess labour out of the fishery and to prevent additional labour from entering it. However, there was no consensus on the criteria defining 'excess labour' – that is, on how to determine who was to be expelled. MacKenzie (1979) argued that the fishery serves as the employer of last resort, with the consequence that excess labour enters the fishery when work is unavailable elsewhere. In a delightful turn of phrase, he declared, 'The received image of the poor fisherman is to be stood on its head – he is a fisherman because he is poor, not the other way around' (816). For MacKenzie, measures were needed 'to eliminate dabblers' (817), and he pointed to growing 'resistance to the influx of newcomers, regarded as interlopers and "moonlighters"' (815). Indeed, MacKenzie argued that the occupational pluralism practised by many fishermen was a response to the fact that they were unable to make an adequate living from fishing alone (812). He saw the elimination of 'dabblers' and 'moonlighters' as one way of establishing a full-time fishery labour force.

In contrast, Crutchfield (1979) argued that the part-timer problem was not a simple one, because 'there are obviously many fisheries in which part-time participation is dictated by the availability of fish, weather conditions on the grounds, or concentrations of fish sufficiently dense to harvest them economically' (748). Noting that most analysts tended to assume that 'economic efficiency would be improved if the fishery were to shift more and more to professional, full-time fishermen' (748), he suggested that licensing regulations could actually hamper such a shift, because fishermen might not have the licences that would enable them to move from one species to another. As a consequence, licensing and limited entry might actually militate against the development of a full-time, professional fishery labour force.[10]

In a similar vein, Scott (1979, 731) chastised those who argued for the expulsion of part-time fishermen on the grounds that they contributed to economic inefficiency. He formulated his objections this way: 'There are many conceivable alternative discriminatory systems: entry can be rationed by race, colour, creed, etc.; by bribery of officials; by queuing; and by lottery. The arbitrary expulsion of part-time and "sport" fishermen with low catches ... should take a prize for high-handed, inefficient discrimination.' Scott, in fact, focused as much on the social aspects of fishery regulation as on its economic aspects. He emphasized that regulation has a 'distributive bias,' by virtue of its 'effectively excluding potential fishermen from one social or economic grouping and conserving or protecting the stock for exploitation by another' (726). He cautioned economists to consider their own motives for regulation (727–8). Scott also warned that there is a clear tendency for fisheries regulations to 'multiply.' 'Overfishing regulations that reduce one component of fishing effort induce further controls to suppress increases in other components' (728). He pointed out that, as a result, 'the net benefit of regulation can easily tend to zero' (728). Indeed, he suggested that the most important thing about fishery regulation that had been learned to date was that ad hoc restrictions were more inefficient than 'the evils of common property' (726).

The third theme apparent in these articles concerns the means of regulating the fishery. Fishery economists weighed the relative merits of regulating the fishery through taxes on excess production or through some form of property rights that would regulate access. From the perspective of classical economic theory, a system of taxes that provided a 'disincentive' for persons to overfish had a distinct advantage. Such a system would allow the fishery to be 'left entirely to the market without fear of biological depletion, excessive inputs in general, or incorrect combinations of inputs' (Crutchfield 1979, 742). There was a general consensus, however, that some form of property rights was preferable to taxation on administrative and political grounds. These economists themselves argued that, as a tool for fishery regulation, a taxation system was simply too open to the possibility of political corruption (Scott 1979, 739). Furthermore, there would be enormous difficulties in getting politicians to risk their political careers by agreeing to accept a level of taxation determined by economists and biologists (Crutchfield 1979, 744–5).

Fourth, all these works assumed that *some* form of 'limited entry' regulation within the fishery was necessary in order to avoid the 'tragedy of the commons'; it was not clear, however, *what* form such regulation should take. Were the conditions of limited entry and conservation best served through the licensing of boats, the licensing of fishermen, and/or the establishment of quotas, that is, 'quantitative rights' (Moloney and Pearse 1979)?

In the 1970s, when the symposium papers under discussion were being written, some limited-entry procedures had already been enacted in both Alaska and British Columbia. Indeed, two of the papers (Fraser 1979, Pearse and Wilen 1979) dealt directly with the West Coast experience.[11] On the West Coast, limited entry was regulated primarily through a system of boat licences and fishing seasons. Licences were related to the size of the boat and were treated as a form of private property that could be bought and sold by individual fishermen. The market value, or selling price, of a licence tended to reflect the total long-term income that the licence could generate. If a fisherman wished to increase the size of his boat, he was required to purchase additional licences for smaller boats equivalent to the length of the new boat.

Under this system, the original licence holders were usually able to make massive personal gains by selling their licences and moving out of the fishery. However, those who purchased the licences were then required to make enough income from them to offset the debt they had incurred in purchasing them. Quite often, in order to do so, they bought larger and more efficient boats and equipment, again increasing their debt level. Such moves placed a further strain on the capacity of the existing fish stocks, with the result that the government often had to shorten the fishing season in which these boats were permitted to operate. This, in turn, could prompt the purchase of even more technologically advanced equipment that would enable the fishermen to catch as much of the fish as possible during the time available. And so the vicious cycle continued.

Primarily because of their awareness of the problems that characterized the West Coast experience, the authors of the symposium papers were generally inclined to prefer some form of quota system (Wilen 1979, Scott 1979), to which they referred as 'quantitative rights.' Moloney and Pearse (1979, 865) pointed out that

the appeal of this approach lies mainly in two unique features, one relating to efficiency, the other to distribution. For the first, quantitative rights cut to the basic cause of economic waste in fishing: rights to take specific quantities of fish largely eliminate individual fishermen's incentives to protect and increase their shares of the catch by defensively and competitively increasing their fishing power. For the second, this technique admits full flexibility with respect to the division of resource rents between the government and the participating fishermen.

The fifth common theme of the works under discussion is reflected in their constant use of the word *rational* to describe the perspective they advocate. To be sure, it may be in the nature of advocacy to suggest that the perspective being advanced is in some way more rational than the competing approaches, or to imply that the latter are in some way less than rational. But the repeated use of the term and the frequent call for 'rational fisheries management' strikes the reader as an extreme example of this tendency. Alternatively, since one way of identifying newly emerging perspectives is by the code words or phrases they employ,[12] the word *rational* and the constant call for 'rational fisheries management' in these papers might be viewed as code words identifying positions that adhere to the 'tragedy of the commons' perspective on fisheries policy.

The authors of the symposium papers were leading fisheries economists and senior policy analysts employed by the Canadian Department of Fisheries and Oceans (DFO). They provided the basic ideas and arguments in favour of the regulation and 'rationalization' of the East Coast fishery. But ideas, even if they are promoted by senior government bureaucrats, do not necessarily become policy – no matter how 'rational' they may appear. Before that happens, the politicians in power must be persuaded not only of the legitimacy of the ideas in their own right but also of the probability that any new policies based on them will not seriously undermine their political base. The problem with the proposals for limiting access to the fishery based on the common-property perspective was that they clearly had the potential to create damaging political repercussions: They would bring about the elimination of a significant number of part-time jobs and put an end to the fishery's function as an employer of last resort.

Thus, the policy process that led up to the implementation of limited-access regulations for the Newfoundland fishery involved a lengthy political debate in addition to the scholarly one that we have just outlined. In its public aspect, it took place in a series of position papers, commission reports, and public addresses by the politicians themselves. It was also characterized by a strong conflict in views between federal and provincial administrators and politicians. It is to the political debate that we now turn our attention.

Acceptance of the Common-Property Perspective by the Federal Government

The political process leading up to the implementation of limited-access regulations to control the Newfoundland inshore fishery involved a number of 'political actors' – groups and individuals at both the federal and the provincial level who attempted in various ways to influence the fisheries policy process and its outcome.

At the federal level there were administrators and advisers who were committed to 'rationalizing' the fishery. There were also, of course, the federal politicians themselves, who, for the most part, had little interest in the question of fishery development. The East Coast fishery is far removed from the country's dominant population bases and, hence, from its dominant areas of federal political support. Thus, a federal decision with regard to the East Coast fishery has relatively little impact on the re-election chances of the party in power. This geographic and political distance sometimes allows the federal government to take what might appear to be a non-political stand with regard to the fishery, or at least one that does not support the dominant-class interests in the Atlantic region. In fact, the federal government has often supported fisheries policies that favour the fishermen over the dominant corporate interests in the fishing industry (Matthews 1983, 193–215). At other times, however, the same distance has given federal politicians a somewhat different sense of political freedom – freedom to implement policies that favour one group of fishermen over another.

An important federal actor in the policy process is the minister of fisheries and oceans. Although the majority of his political colleagues may be far removed from local issues on the East Coast, the minister himself is invariably from one of the four Atlantic provinces or British Columbia, provinces in which the fishery is a dominant industry. Throughout the period from the mid-1960s to 1984, all but one of the federal fisheries ministers were from the East Coast and thus tended to have a personal interest in the potential social and political impact of limited-access regulations on that region. A particularly important figure was Roméo LeBlanc, a New Brunswick native who served as fisheries minister during the late 1970s and early 1980s and who played a direct, decision-making role in the development of Canadian fisheries policy.

Although the federal government had instituted a licensing and 'rationalization' program in the British Columbia fisheries as early as 1968 (Copes 1980, 136–9; Draper 1981, 118), no similar program was introduced in the East Coast fisheries at that time. Federal efforts on the East Coast were being focused on a different project. In 1973 (and again in 1974) the Canadian East Coast fishery underwent virtual economic collapse (McCorquodale 1983, 159). Between 1974 and 1976 the government found itself having to provide more than $140 million in additional subsidies to fishermen and fish-processing plants (Kirby Task Force 1982, 19). In response to the crisis in 1973, at the Law of the Sea conference, Canada attempted to gain international consent to extend its jurisdiction over fisheries to 200 nautical miles from its coastline. When this was slow in coming, Canada threatened to declare a 200-mile fishing limit unilaterally. Ultimately, the multinational conference granted Canada the right to extend its jurisdiction over fisheries to 200 miles effective 1 January 1977 (Copes 1980, 141–2; Draper 1981, 120–1).

Also in 1973, Canada announced that it would introduce a general licensing system for East Coast fishermen. (The system would become fully operational only in 1975). In addition to fishermen's licences, the system would now require licences for boats (of all sizes) as well. Licensing of lobster fishermen had been inaugurated on the East Coast as early as 1967, and this type of program had been quickly extended as a means of controlling the catch of other shellfish, as well as salmon and herring (Sinclair 1981, 19). However, there had been no previous attempts to restrict the catch of groundfish, such as cod,[13] or to require any form of general licence or boat licence for fishermen.

The impetus for licensing was the federal government's expectation that its 200-mile fishery jurisdiction would allow it to set quotas and regulate catches throughout the whole of the East Coast fishery (Sinclair 1981, 10). To do that, it would need a detailed inventory of fishermen. Yet, in announcing the plan, the then fisheries minister, Jack Davis, declared that the federal initiative was a response to 'a sudden upsurge in Atlantic fishing vessel construction in 1973' and that it was meant to encourage expansion of the offshore sector. He also stated that such a licensing program would not harm any current fishermen, suggesting that it was simply a measure to identify legitimate fishermen: 'All fishermen presently in the industry will be protected ... To do this we need to know who are bona fide fishermen and who are not' (Canada, House of Commons 1973, 7775–6).

To a considerable degree, the licensing policy inaugurated in 1974–5 was consistent with the minister's 1973 announcement. The licensing system operated simply as a registry, and no distinction was made between full- and part-time fishermen. Moreover, there appears to have been no attempt to use the licensing system as a vehicle for barring from the fishery those who would use it as an employer of last resort. Licences were freely available to all who requested them, as is evidenced by the considerable growth in numbers of licensed fishing vessels and officially registered fishermen during the period 1975–81.[14] In short, though licensing of inshore fishermen was inaugurated in Newfoundland in the mid-1970s, it operated simply as a registration program. There was no attempt at that time to use it as a means of limiting access; in other words, public policy still had not addressed the problem of 'the tragedy of the commons.'

The first public recognition by the federal government of 'the tragedy of the commons' on the East Coast is to be found in a May 1976 document entitled *Policy for Canada's Commercial Fisheries* (Canada, Environment Canada 1976). In a discussion of the Atlantic fishery it declared that 'in an open-access, free-for-all fishery, competing fishermen try to catch all the fish available to them, regardless of the consequences. Unless they are checked, the usual consequence is a collapse of the fishery: that is, resource extinction in the commercial sense, repeating in a fishery context "the tragedy of the commons"' (39). The report recommended a reduction in the number of fishermen employed in the primary fisheries in Atlantic Canada, but warned that such a reduction 'would have a different effect in different communities ... Where adverse social side effects such as reduced employment opportunities can be kept within acceptable limits, restructuring should proceed' (56). As Copes (1980, 142) notes, 'The document clearly confirmed the economic analysis that had emerged over the previous fifteen years. It acknowledged the need to apply limited entry universally, to reduce significantly the excessive manpower of the inshore fishery, and to rationalize the dispersed and fragmented processing industry.' Although the document did not so much 'confirm' the economic analysis as assert it, Copes is right in observing that the significance of the 1976 policy document lay in its explicit acknowledgment and acceptance of the common-property perspective developed by economists and its declaration that the way to overcome such a problem is through a reduction in the number of fishermen.

Although the policy document presumably outlined the policy position of the federal government with regard to licensing, it did not clearly articulate how the government would deal with two key issues: the relationship of the inshore fishery to the offshore fishery and the relationship between full-time and part-time fishermen.

With respect to the first issue, Roméo LeBlanc, who replaced Jack Davis as minister of fisheries and oceans, was clearly instrumental in promoting the interests of the inshore and nearshore fishermen over those of the offshore fishermen (Copes 1980, 143–4; Matthews 1980; 1983, 194–215). One reason that the federal government did not clearly articulate its stance on part-time fishermen in the 1976 policy document or in subsequent announcements may have been the potential political consequences of doing so. In fact, there was a deliberate effort in subsequent federal statements to imply that, in some unspecified way, limiting access through licensing and quotas would benefit all fishermen. Thus, on one occasion, the minister referred to licensing as 'regulating for people' (LeBlanc 1978a, 2) and, on another, claimed that 'by means such as limited entry, licence control, quotas and overall fleet coordination, we protect for each fisherman his share of the fishing grounds. Our first preoccupation is to protect existing fishermen and consolidate the good health of their fishery' (LeBlanc 1978b, 3).

When it was impossible to avoid the part-timer question, the federal government chose to retain the euphemism it had used when the policy was first announced in 1973 – *bona fide fishermen.* Just as the term *rational* became a code word to identify economic theorists who espoused the 'tragedy of the commons' perspective, so the term *bona fide fishermen* became a code phrase used by policy makers who were increasingly committed to licensing as a means of limiting access to the inshore fishery. Thus, in a highly confrontational speech to the fisheries ministers of the Atlantic provinces, the federal minister asked, 'When the increase in cod trap fishermen means the bona fide fisherman, by the luck of the draw, finds his berth in a barren patch, will the province encourage restrictions on the numbers of new fishermen?' (LeBlanc 1978a, 14). Similarly, a DFO news release on the effects the licensing regulations might have on fishermen was quick to emphasize that 'bona fide fishermen' would not suffer: 'On the whole Atlantic coast there will be, however, no restrictions on bona fide fishermen, as determined through an appeal committee, using longline and baited trawls, and on non–bona fide fishermen using handlines' (Canada, DFO 1979: 2). The obvious implication of the statement was that some 'non–bona fide fishermen' would be subject to considerable restriction. The only fishermen who seemed to fit the circumlocutory category of 'non–bona fide fishermen' were those who fished part time using gear that was more sophisticated than the simple handline. From the fishermen's perspective, the announcement essentially confirmed that government fisheries policy would henceforth involve discrimination against the vast majority of part-time fishermen.

The DFO news release reflected a decided shift in federal licensing policy, beginning around 1979, that was undoubtedly influenced by the massive increase in the number of licensed fishermen. The immediate impetus for the shift, however, was a review of Canadian fishery policy undertaken by C.R. Levelton of the DFO at the request of the minister, Roméo LeBlanc.[15] Levelton's instructions explicitly required him to consider the role of licensing in the groundfish fishery. He was to 'review and evaluate the licensing systems of Canada's east coast commercial fisheries and provide recommendations concerning the role and type of a future licensing and fee system. Particular emphasis will be placed on the groundfish fishery and its relationships with other fisheries' (Levelton 1981, 2). Levelton's final report provided a long and detailed list of recommendations. The following four directly affected inshore fishermen: (1) that licences be issued to individuals rather than to vessels; (2) that a categorization of licences be implemented in order to differentiate among 'regular fishermen, apprentices and casual fishermen'; (3) that the sale and transfer of licences from one person to another be prohibited; and (4) that there be a simple and universal registration of all fishing vessels (Levelton 1981, 83–5).

The minister revealed his position on the Levelton report in a March 1980 speech to the United Maritime Fishermen. He stated: 'we need further definition of who should get a chance to fish, who should get a licence ... It's time we dusted off the Levelton Report and made some decisions on such matters. Meanwhile, until the existing fleet has enough fish for a decent living, and until we think through the whole question, we should as a rule give out no additional licences' (LeBlanc 1980, 5). LeBlanc made clear his opposition to both 'the rationalizers wanting to develop and consolidate everything in sight, even if it means doing away with small fishing villages,' and 'the rural romantics arguing that we must preserve the coastal way of life at all costs, even when it means preserving poverty' (2). At the same time, however, he betrayed his own acceptance of 'the rationalizers,' position in this eloquent restatement of the 'tragedy of the commons' argument: 'If you let loose that kind of economic self-interest in fisheries, with everybody fishing as he wants, taking from a resource that belongs to no individual, you end up destroying your neighbour and yourself. In free fisheries, good times create bad times, attracting more and more boats to chase fewer and fewer fish, producing less and less money to divide among more and more people' (3). In retrospect, such a statement would suggest that official acceptance of limited-access licensing could not be far off.

LeBlanc made the first public announcement of federal acceptance of such a policy only six months later, in September 1980, in his closing remarks to a conference in Memramcook, New Brunswick on the groundfish fishery.[16] After describing how he had listened to all the deliberations of the conference participants and stayed up late into the evening reading their submissions, he informed his audience that he was now presenting his own final decision because someone in his position 'can't pass the buck to somebody else, and at some point what he has heard will have to be either accepted or rejected, or somewhat adjusted' (LeBlanc 1980a: 2). He even went so far as to suggest that, in forming his views, he had rejected the position laid down for him by his own officials: 'If you detect a certain nervousness amongst some of my officials, it's because I decided the speech that was written for me before the seminar would not be very convincing, if I heard something new and different at the seminar. So for that reason I'm speaking from my own handwriting, and if sometimes I stumble, it has to do more with my writing than my elocution' (2).

It is unlikely that LeBlanc's advisers were upset with the announcement that he then went on to make – namely, that limited-access licensing involving a distinction between full-time and part-time fishermen would be introduced early in 1981. However, it was a mark of LeBlanc's political skill that he managed to make this announcement in such a way as to imply that it was contrary to the wishes of his officials, and was, rather, a direct response to the requests of '25,000 fishermen.' Furthermore, he left the impression that the administration of the program would be largely in the hands of the fishermen. The following passage from LeBlanc's statement (1980a, 5–6) conveys the flavour of his announcement, and outlines some of the details of the ensuing licensing regulations:

The fact is that we have been talking about licensing long enough, and we must move ... Licensing is a chief concern of the industry as a whole, but especially of fishermen ... On the basis of very extensive and exhaustive consultations that have already been held with more than 25,000 fishermen and other representatives of the industry, it looks as if a consensus is emerging, or has emerged, in favour of categorizing personal fishing licences into three main groups – the full-time, the apprentice and the part-time or casual. I want to make it very clear we are not proposing to threaten anybody's modest livelihood. We're looking at categories of fishing licences.

Coupled with such a new regime will be the creation of local licensing allocation and appeal committees chaired by people who are not members of the federal bureaucracy. The majority of the members of these committees will be fishermen. My intention is to implement these two key recommendations early in 1981 ...

These two additional changes alone should go a long way in recognizing legitimate local concerns in the management of the fishery, by giving fishermen a direct voice in licensing issues which affect their livelihoods and their lives. So that there will be no misunderstanding, I want to make it clear that these are the kinds of changes being proposed and requested by the fishermen themselves, and not by some committee of officials in some supposed ivory tower.

Thus, the shift in the value orientation of fishery regulation, based on economic theory of common property, finally became a matter of state policy in 1981. Moreover, it would appear that LeBlanc played a critical role in determining the direction and shape of fisheries policy.

The Newfoundland Government's Developing Opposition to Limited-Access Licensing

To this point, our discussion has focused solely on the acceptance of the 'tragedy of the commons' perspective by the federal government. But the policy process leading to the implementation of limited-access regulations for the Newfoundland fishery involved a lengthy political debate that ultimately led to an open conflict between federal and provincial administrators and politicians.

Newfoundland's position on common property and limited-access licensing can best be understood in the context of the province's historically distinctive approach to fishery regulation, as well as of its broader social policies, such as the Newfoundland Resettlement Program, which had a major impact on the communities in which fishermen live.

Although most treatments of East Coast fishery policy start with the British North American Act of 1867, it is important to an understanding of Newfoundland's position on limited-access licensing to bear in mind that Newfoundland did not join Canada until 1949 and that its previous history of fishery regulation was characterized by a distinct value orientation. Martin (1973, 1979) is the only writer on the subject who focuses on the value orientation of Newfoundland's fishery policy prior to Confederation. He traces the development of fishery regulation in Newfoundland to an 1890 act of the General Assembly of Newfoundland. From his description there appear to have been significant differences in purpose and value orientation between Newfoundland's and Canada's fisheries policy.

First, while early Canadian policy emphasized the importance of conservation (Draper 1981, 114; McCorquodale 1983, 156), according to Martin (1979, 284), Newfoundland's regulatory efforts 'were not enacted primarily with a view toward husbanding the resource, i.e., conservation. Rather they are [sic] a political response on the part of the government to the ecologic and social needs of various inshore fishing communities.' Martin contends that the Newfoundland government originally became involved in fisheries regulation because of the pressure brought to bear on elected representatives by inshore fishermen demanding that something be done to settle local fishing disputes arising out of the use of different technologies (184–6). Hence, Newfoundland's traditional fishing regulations were social policies designed to maintain a status quo more than they were conservation policies. Such regulations allowed a local fisherman to 'remain secure in the knowledge that he would have at least an equal chance (ideally) to catch his rightful share of the resource' (298).

Second, Martin argues that Newfoundland fisheries regulations supported the cultural façade of equality and egalitarianism that was maintained in rural Newfoundland fishing culture. He contends that Newfoundland's regulations were essentially the codification of local arrangements into law, thereby permitting 'the avoidance of conflict and the maintenance of the egalitarian-relations veneer' (292). The regulations were designed to ensure that actual relations on the fishing grounds, which involved 'distrust, competition, and manoeuvring,' were 'not allowed public acknowledgement ashore' (289). Rather than attempting to maximize the benefits from fishing of any one category of fishermen, Newfoundland fishing regulations seem to have been directed towards maintaining both the equality of opportunity and the equality of results.

Martin's depiction of early fisheries polices in Newfoundland as being designed to preserve the egalitarian and non-competitive character of Newfoundland fishing villages rings true, given the character of life in most isolated rural Newfoundland communities. Indeed, under the conditions that prevailed there, with people living and working in close proximity and isolation while competing for the same resource, it seems reasonable to assume that numerous cultural proscriptions and norms designed to defuse the obvious basis for potential conflict would develop. Whether, in reality, fisheries regulations and norms were oriented this way in Newfoundland more than in Canada's Maritime provinces is open to question, though there is nothing in the literature on the development of Canadian fisheries policies that in any way suggests a concern with egalitarian considerations or with the regulation of community conflict. Given this, and based on Martin's analysis, it seems likely that the value orientation of Newfoundland's fisheries policy was different from that of Maritime Canada's.

If so, Newfoundland fishermen today may view the role of the state in the regulation of the fishery somewhat differently than do fishermen elsewhere in eastern Canada. Based on historical precedent, they might believe that the role of the state is to support their traditional practices rather than to provide a regulatory system based on a different set of values. They may well see the role of state regulation as reinforcing the veneer of equality among fishermen. However, as we have seen, the values inherent in current Canadian fisheries policy give little credence to the notion that local customary regulations can police the fishery adequately. Moreover, current federal policy attempts to enhance the economic benefits received by some fishermen, even if it thereby reduces the benefits received by others.

To be sure, it might be argued that events since Confederation have done much to undermine Newfoundland fishermen's belief in their provincial government's commitment to maintaining equal rights and the status quo. For example, the Newfoundland Resettlement Program was a deliberately orchestrated federal and provincial attempt to undermine the traditional, communal way of life and the values on which it was based (Iverson and Matthews 1968, Matthews 1987). The vehemence of popular oppositions to such policies ultimately became so great, however, that the government was forced to abandon them. Indeed, by the early 1970s, it was virtually impossible for any Newfoundland administration to openly advocate policies and programs that would in any way undermine the traditional values and the integrity of rural community life. In that sense, the resettlement program, while being a clear exception to the generalization that Newfoundland fisheries policies emphasize equality and egalitarian norms aimed at reducing communal conflict, may in fact be the exception that re-established the rule. By the early 1970s the Newfoundland government had to be careful that it was not seen as supporting policies that could in anyway be construed as promoting the abandonment of rural fishing villages.

In examining Newfoundland's response to federal limited-access policies, it is necessary to consider a number of actors, both within and outside the provincial government, who were involved in various ways in shaping the policy process and the provincial response. The provincial politicians and their advisers played a major role in the debate, but their interests in the matter were not necessarily undivided. For example, given the persuasiveness of the economic arguments, they might be expected to favour the development of a strong professional labour force of full-time fishermen. That position, however, implied the likelihood of hardship for many part-time fishermen, and those part-time fishermen were also voters. Consequently, the provincial government's reaction to the imposition of licensing and limited-access policies might best be described as mixed.

The initial provincial reaction was a general, if somewhat guarded, approval of licensing and limited-access. Hence, in 1973, only two days after the federal fisheries minister declared his intention to introduce a licensing scheme for the East Coast fishery, his Newfoundland counterpart, Harold Collins, publicly expressed his support for the action. He indicated that this policy would satisfy 'the need to match the fishing efforts [to the existing] resources,' and assured fishermen that there was no intention in such action of 'plotting the destruction of the Newfoundland inshore fishery' (quoted in Committee on Federal Licensing Policy 1974, 14).

Although such statements indicate support for the initial use of licensing as a means of 'registering fishermen,' they do not necessarily indicate support for its use as the basis of a program of limited access. However, in 1978, the provincial government released its *White Paper on Strategies and Programs for Fisheries Development to 1985* (Newfoundland and Labrador, Ministry of Fisheries 1978), which, although it never explicitly mentioned limiting access to part-time fishermen, nevertheless implicitly supported the idea. It stated that the fishery could 'no longer remain an employer of last resort,' that there was a need for a 'select corps of professional fishermen,' and that licensing would benefit those who could 'maintain an effective presence within the harvesting sector':

From a licensing policy perspective, for example, the aim of both levels of government must be to ascertain, with input from fishermen's organizations and industry, that level of fishing effort which each species fishery can sustain in order to generate a reasonable income for those who can maintain an effective presence within the harvesting sector.

Given this approach, it is inevitable that the fishery will no longer remain an employer of last resort. The status of fishermen will rise, since with appropriate levels of training, a select corps of professional fishermen will emerge over the longer term. (21–2)

However, the same document also emphasized the Newfoundland government's continuing support of traditional settlement patterns:

Social and economic considerations are, in the final analysis, the basis around which fisheries development strategies are initiated and implemented ...

The Province's commitment to fisheries development and its commitment to maintaining the settlement pattern which exists throughout Newfoundland and Labrador reflects the contribution which the fishing industry makes to the economy both from a social and an economic perspective. (21–2)

The problem not addressed in this policy document was that the two positions were largely incompatible. Fisheries-development policies based on economic considerations involved the closure of the fishery as an employer of last resort. Such policies conflicted, at least to some degree, with social policies committed to the maintenance of the traditional settlement pattern.[17]

Nevertheless, between 1978 and 1980 the Newfoundland government clearly overcame its apparent ambivalence, and ultimately opted to support the social value of an open fishery over the perceived need to close the fishery as an employer of last resort. A number of factors may have contributed to this decision. One was the rise of a significant level of local opposition. For example, a group of university professors had formed the Committee on Federal Licensing Policy[18] and had issued a private report (1974) cautioning against the use of licensing as a means of excluding part-time fishermen from active involvement in the fishery. The report argued against licensing on two grounds. First, it contended that 'the proposed licensing policy seems to rest on the assumption that the provincial government and local fishermen do not already have the necessary means to control entry and allocation to the inshore fisheries without resorting to a higher organizational level.' It then pointed out that 'entry to the inshore fishery, far from being wholly uncontrolled (as is often assumed), has long been regulated according to customary rules and regulations emanating from the local level' (18). In short, it challenged the very 'tragedy of the commons' argument on which the limited-entry policy was based. Second, the report pointed out that, historically, the survival of many coastal communities, particularly along the northeast coast of Newfoundland, had depended on 'occupational pluralism.' It argued that limited-access licensing directly threatened the survival of many of them: 'In Newfoundland today it [occupational pluralism] is still an adaptation to a shortage of employment opportunities which are not seasonally based. Any fisheries licensing policy which ignores that fact is on dangerous ground. It should be understood that implied in all this is the staggering cost of relocating a substantial proportion of the remaining rural population of Newfoundland and Labrador to urban centres' (22). As noted above, even by the early 1970s the Newfoundland government had become particularly sensitive to any accusation that it was engaged in a program involving rural relocation.

The province's position in 1978 was contained not only in the White Paper but also in an address by the premier of Newfoundland, Frank Moores, to the national First Ministers Conference, in February.[19] Moores (1978, 5) argued that responsibility for all fishery licensing and the allocation of fishery quotas should be turned over to the province: 'I now call upon this Conference to support the principle that control of licensing policy be delegated to the Province for a five-year period, and that the Federal Government recognize our right to participate in the establishment and allocation of quotas.' Then, in August 1980, the Newfoundland government established a royal commission with the mandate to inquire into 'the impact of limited entry and other licensing schemes upon the incomes of inshore fishermen, the future evolution of the inshore fishery, and the social and economic development of communities predominantly dependent upon the inshore fishery' (Paddock Royal Commission 1981, x). However, the commission had barely been appointed when federal fisheries minister Roméo LeBlanc announced his decision to implement a limited-access licensing policy.

The other major factor contributing to the growing provincial opposition was the election of Brian Peckford as premier of Newfoundland. Peckford had campaigned largely on a 'nationalist' platform of self-sufficiency and opposition to domination by the federal government in Ottawa. He was clearly outraged at the federal government's decision to implement limited-access licensing, at its rejection of the province's request for jurisdiction over licensing, and at its decision to proceed without waiting for the provincial royal commission to report. Thus, in the fall of 1980, shortly after LeBlanc's announcement, Peckford's new government issued a pamphlet that strongly opposed the federal licensing program. It read, in part, as follows:

A new licensing policy has been introduced by the Federal Government which in effect classifies $^2/_3$ of our fishermen as 'part-time.' This could have a tremendous effect on hundreds of small rural communities that depend in whole or in part on the inshore fishery.

Along with the new licensing policy, the notion of an inshore quota has been introduced. We can see the day coming when the inshore fishing could close down in late August due to quota restriction. What happens to the poor fisherman who had a bad July and wants to make it up in September?

Your provincial government does not agree with these policies or the abrupt way that they were put in place. (Newfoundland and Labrador, Ministry of Fisheries 1980, 3)

The document concluded with the statement that 'major decisions should await the outcome of the Royal Commission and the constitutional process' (4).

The Newfoundland government under Brian Peckford continued to maintain its strong opposition to the federal licensing program with its provision for limited access. In 1982 it issued a statement declaring that 'the Provincial Government has taken the view that the right to fish is a local birthright ... We would put limitations on the number of larger boats and the amount of gear, not the number of fishermen' (quoted in Copes 1983, 26). Thus, the province might be described as having opted for 'social' and 'settlement' considerations over those related to economic rationalization.

From the perspective of political economists who would argue that the state operates in the interest of the dominant capitalist class, the province's position seems to make little sense. In the political and social context of Newfoundland, however, the government had little choice. As Copes (1978, 165) has noted, 'a deliberate rationalization of the fishery that bars a proportion of the fishing population from the only employment available to them is socially and politically unacceptable. For this reason, it appears almost impossible to achieve substantial rationalization of the fishing industry before the general employment situation ... has been improved.'

The value basis of Newfoundland's provincial fisheries policy today thus remains very similar to that of its pre-Confederation policies as described by Martin (1973, 1979). The concern of fisheries policy still rests primarily with its social implications rather than with conservation issues. Furthermore, there is a clear interest in maintaining the egalitarian ethos that has traditionally been a major – if largely mythical – part of the rural Newfoundland value system.

The Role of the Newfoundland Fishermen, Food and Allied Workers Union in the Implementation of Limited-Access Licensing

In examining the implementation of limited-access licensing policies in Newfoundland, we must consider the role of one other actor on the provincial level. This is the Newfoundland Fishermen, Food and Allied

Workers Union (NFFAWU), which represents all fishermen – inshore, nearshore and offshore – whether part-time or full-time. Given the conflicting interests of these various groups of fishermen, particularly with respect to limited-access licensing, the stance of the union on the issues of licensing and limited access became a critical factor.

The NFFAWU has often been faced with the difficult choice of having to support the interests of some of its members at the expense of the interests of others (Matthews 1983, 202–4). Limited-access licensing forced the union to choose between the interests of some 8000 members who were full-time fishermen and those of some 15,000 members who were part-time fishermen. The union chose to serve the interests of full-time fishermen.

In fact, the NFFAWU had long been an ardent advocate of limited-access licensing and had lobbied the federal government to have it implemented. In 1977, even before an official licensing distinction between full- and part-time fishermen was introduced, the union had incorporated such a distinction into its contract with fish-processing facilities. Under the agreement, in times of a 'fish-glut,'[20] fish buyers were required to purchase from 'bona fide fishermen' before turning to 'moonlighters.' The agreement also established that the union local in any community was the body that would determine who the 'bona fide fishermen' were. The agreement (NFFAWU 1977, 9) was explained to fishermen in a union announcement that read, in part, as follows:

The key new clause in the contract deals with the handling of fish during the June glut.

'In each area', says Article 4.06 of the contract, 'a fisherman's committee will submit a list of bona fide fishermen to the companies concerned, to enable the parties to ensure that these fishermen have the first opportunity to sell their catch to those companies they regularly supply, during periods of over-supply.'

What this means is that during the glut, the bona fide fishermen will get priority over moonlighters in the sale of fish, provided the fishermen's committee lists only full-time fishermen. If the committee wants to include moonlighters on the list it can do so, but it will be at the expense of the full-time fishermen ...

There's an old saying in Newfoundland that 'everyone has a right to fish', but what the new collective agreement establishes is that full-time fishermen enjoy special rights in selling their catch.[21]

In the same vein, when the federal government appointed C.R.

Levelton to undertake a review of licensing, the union submitted a proposal to him stating that it 'recognizes controlled entry as one of the key tools of fisheries management.' The NFFAWU submission also proposed a division of licence holders into two categories, 'A' and 'B.' 'B'- category licence holders would have the right to fish, but would not be allowed to sell fish during a glut or to participate in the community 'draw' for trap berths (NFFAWU 1979b, 14). Levelton supported the union proposal for limited access, and the union applauded his report. It also described what it considered to be the 'two main parts of the licensing problem': 'First, the question of who are the bona fide fishermen who should get first crack on the grounds, financial assistance and so forth; and secondly, the question of how access to the restricted fisheries (lobster, salmon, herring, crab, shrimp) should be distributed among these bona fide fishermen' (NFFAWU 1979a, 13). In making such a statement, the union was 'upping the ante' by implying that protected-species licences now held by part-time fishermen should be taken from them and awarded to 'bona fide,' full-time fishermen.

By March 1980 the union was publicly giving the federal government advice about other aspects of licensing policy as well. It argued that the government should 'licence the man, not the boat' and that the licence should revert to the government when the licensed fisherman 'ceases fishing' (NFFAWU 1980a, 21). In April of that year the union praised the statement of the DFO representative appointed to deal with licensing in eastern Canada that 'a committee of fishermen should tell the federal government who the bona fide fishermen are ... instead of the federal government telling the fishermen who are the bona fide fishermen' (NFFAWU 1980b, 18).

It was undoubtedly these clear statements by the union that formed the basis for fisheries minister LeBlanc's claim that, in deciding to implement a limited-access licensing policy, he was responding to the requests of 25,000 fishermen: The NFFAWU, after all, represented roughly that many fishermen. It was even more significant that the policy announced by LeBlanc contained almost every feature that had been advocated by the union: (1) it distinguished between full-time and part-time fishermen; (2) it licensed the man, not the boat; (3) it required that licences revert to the government when not in use, rather than becoming private property; and (4) it declared that the union's fishermen's committee in each community would have the power of decision in all appeals of part-time status.

The acquisition of final decision-making power over the allocation of full-time versus part-time licences was a massive and strategically important victory for the NFFAWU. Moreover, it was an absolute defeat for the Newfoundland government. In formulating its policy, the federal government had ignored the province's request for control over the licensing of fishermen, and had instead delegated a significant portion of that power to the union. In 1978 Premier Moores (1978, 3) had clearly described the powerlessness of the Newfoundland government, as follows:

Newfoundland has a smaller measure of control over our economic destiny than any other Province in this Federation. We believe that we must have greater control over our energy and fishery resources ... In the case of fisheries, we as a Province have neither ownership nor control over a resource which is of vital importance, socially and economically ... I don't think there is another Province in Canada whose economic sovereignty is as seriously impaired as is that of my Province.

Then, in 1981, the province was forced to witness the granting to a third party of a significant measure of the control that it so desperately wanted over one of its key resources. Under the circumstances, it is easy to understand the anger and frustration that Premier Peckford and other members of his administration displayed in their response to the federal licensing program.

By 1981, Newfoundland had lost two battles over the fishery. The first pertained to the definition of the problem and the desirable direction for further policy. The province had been unable to convince the federal government that a concern for 'social considerations' should take precedence over the considerations of economic rationalization. However, even more insulting was the loss of the second battle, for control over the policy-implementation process. The federal rejection of Newfoundland's plea for some measure of control over the harvesting of one of its major resources, and the alliance between the DFO and the NFFAWU with regard to the regulation of licensing, virtually guaranteed a continuing conflict between the two levels of government responsible for bringing 'rationality' to the fishery.

Conclusion

The implementation of limited-access licensing in the Newfoundland inshore fishery provides the policy context for this study. The acceptance of a policy of limited-access licensing has been shown to have occurred on two levels. At the level of ideas and metaphor, there was the development of the 'tragedy of the commons' perspective, which ultimately brought about an acceptance of the notion that economic principles should be at least as important as biological conservation in determining the nature of fisheries policy. Acceptance at this level made possible the transformation of public administration that allowed limited-access licensing to become a reality.

Throughout this process diverse social and economic interests were championed by the various political actors involved. We have shown that at the root of the conflict between the federal and the provincial government was the extent to which each subscribed to the principles of economic rationalization or to values that gave a priority to 'social and settlement considerations.' (Of course, each government's willingness to espouse the latter was directly related to its own political interests.) We have also suggested that Newfoundland's current policy in this regard reflects the continuation of its pre-Confederation value orientation. Finally, we have attempted to document the growth of an alliance between the federal fisheries administrators and the Newfoundland Fishermen, Food and Allied Workers Union. As we have seen, the federal government delegated to the union a significant area of responsibility in the administration of the limited-access licensing regulations. In doing so, it ignored the provincial government's calls for greater control over one of its major resources.

In sum, this chapter has demonstrated that, by 1981, the federal and provincial governments had both made clear the values and rhetoric on which their conflicting positions were based. However, the federal government's position, based as it was on established economic theories of common property, had a clearly articulated set of assumptions, propositions, and conclusions that contrasted sharply with the less systematically developed statements about maintaining the quality of community life that were basis of the Newfoundland government's position.

It is frequently the case when arguments for economic efficiency are pitted against concerns for social vitality and quality of life that the former carry an aura of objective truth, while the latter seem to be based on largely unsubstantiated assumptions and suggestive beliefs. In presenting the economic theory of common property, to which we refer here as the 'tragedy of the commons' perspective, we have generally refrained from offering a critical assessment, largely because the theory became the basis of Canadian limited-access policy virtually without benefit of critical examination of its intellectual content. Such criticism as was offered tended to focus on the deleterious social consequences of limited-access policies, rather than in any direct way questioning the assumptions of the 'tragedy of the commons' perspective itself.

In the last few years, however, a copious body of literature and research has appeared that directly questions the premises on which the 'tragedy of the commons' argument rests. This body of literature provides considerable empirical evidence of clear alternatives to the role of the state implied in the 'tragedy of the commons' argument. It also provides evidence that, in most local fisheries, open access does not involve the unbridled competition of all against all. Moreover, it emphasizes that limited-access licensing, in itself, does not reduce competition for fish, but simply limits the right to compete to a select number of fishermen. The fishermen who are able to obtain licences remain in competition both for fish and for income to be derived from the sale of fish. This recent body of work demonstrates that the federal and provincial positions on limited-access licensing are in a sense based on competing metaphors, providing alternative visions of the values that should inform fisheries policy. They are the focus of our analysis of the 'conceptual context' in chapter 4.

4 Understanding Common Property
Social and Social-Psychological Dimensions

The meaning of property is not constant. The actual institution, and the way people see it, and hence the meaning they give to the word, all change over time. We shall see that they are changing now. The changes are related to changes in the purposes which society or the dominant classes in society expect the institution of property to serve.

C.B. MacPherson,
'The Meaning of Property' (1978, 1)

The preceding chapter documented the acceptance of the 'tragedy of the commons' perspective as the basis of federal fisheries policy for Canada's East Coast. Attention was given to the very different orientations of the Canadian and Newfoundland governments with regard to the regulation of the fisheries. We depicted those orientations as competing metaphors that expressed different world-views.

Thus far, however, we have focused largely on the policy level. Thus, although we outlined the assumptions of the 'tragedy of the commons' perspective, we did not attempt to locate that perspective in the context of general theory on the nature of property. Similarly, although we documented the Newfoundland government's opposition to the 'tragedy of the commons' perspective from a historical and political point of view, we did not consider whether that opposition might also be understood in the context of existing theories on the nature of property and the role of the state in its regulation.

As noted earlier, opposition to the 'tragedy of the commons' perspective on a more abstract and general level has grown in recent years. Theorists and researchers from a wide range of fields, including political science, geography, psychology, sociology, anthropology, and even economics, have begun to question the fundamental assumptions of the theory, namely, that 'freedom in the commons brings ruin to all' (Hardin 1968, 1244) and that 'under conditions of open access the tendency to overcrowding and depression is universal and inexorable' (MacKenzie 1979, 811). The conceptual basis of the opposing position is drawn primarily from political science theories of public choice and psychological game-theory models. Its empirical foundations are rooted largely in anthropological studies of small-scale societies. The argument is made that, in theory, as long as certain enforcing conditions prevail, the depletion of common-property resources is not an inevitability. This proposition is supported by historical and anthropological evidence of common-property situations in which resource depletion did not occur.

In this chapter, we compare the tenets and assumptions of this alternative perspective on the nature of common property with those of the 'tragedy of the commons' perspective. We focus especially on the extent to which each of the two perspectives embodies different assumptions about the nature of property, in general, and of common property, in particular. In that sense, the analysis in this chapter is more abstract than that found in chapter 3. We also examine the different approaches to the regulation of common property that advocates of the two opposing positions espouse.

To understand correctly the differences between these two positions, we must first examine some broader questions about the nature of all property, as well as of different *types* of property (including common property). It is also important, as part of this consideration, to examine the role of the state in regulating property. This broader perspective on the nature of property in all its forms will aid us in identifying the strengths and weaknesses of different perspectives on the nature of common property. It will also make us better equipped to determine the most appropriate role for the state in regulating common property.

It might be argued, particularly by those who are not enamoured of theoretical discussion, that the level of conceptual analysis just described is of little value here. To be sure, a consideration of the origins, nature, and types of property may seem somewhat removed from the more pragmatic examination of the transformation of the fishery and fishing communities of rural Newfoundland that is our main objective here. We would suggest, however, that a consideration of these broader issues is, in fact, of direct relevance to our understanding of the way in which the Newfoundland inshore fishery has traditionally been regulated, and provides the basis from which we can make an informed assessment of the most appropriate role for government in regulating the fishery today. The opposing theoretical positions highlighted here are reflected in alternative visions of the social vitality and economic viability of fishing communities. Through an examination of such theories and of empirical evidence that their proponents claim support them, we can better understand the present situation in rural Newfoundland.

The Nature and Types of Property

Property, in common parlance as well as its dictionary definition, is something that one owns. Such a definition implies that property is a thing, an object that is possessed. As MacPherson (1978, 3) notes, however, 'as soon as any society, by culture or convention or law, makes a distinction between property and mere physical possession, it has, in effect, defined property as a right.' Property, then, is not so much an object as a relationship (or a set of relationships) whereby an individual or group is accorded the right to use something. It is important to recognize that a *right to use* implies, by definition, a *right to limit access* as well. In other words, property rights and the limitation or restriction of access are alternative sides of the same relationship.

The distinction between the right to use and the right to limit access, or to exclude, is the basis of two important attempts to categorize types of property rights. One, developed by W.P. Welch and used by Ostrom (1987a, 252), distinguishes three types of property: 'Common property can be used by anyone ... Usufruct property confers a nontransferable right to exclude others from use ... Full ownership confers both excludability and transferability.' The other categorization of types of property was devised by MacPherson (1978). Like Welch, MacPherson recognizes the importance of the distinction between the right to use and the right to exclude in identifying categories of property rights. However, he adds a further dimension often ignored by other theorists, namely, the role of the state. As a result, MacPherson distinguishes between state property, private property, and common property, purportedly on the basis on the type of right involved, but with a particular emphasis on the right to limit or exclude. He is particularly concerned with the nature of common-property rights, which he argues are not rights of the state but rather rights of individuals:

The fact that we need some such term as 'common property' to distinguish such rights from the exclusive individual rights which are private property may easily lead to our thinking that such common rights are not individual rights. But they are. They are the property of individuals, not of the state. The state indeed creates and enforces the right which each individual has in the things the state declares to be for common use ... The state *creates* the rights, the individuals *have* the rights. Common property is created by the guarantee to each individual that he will not be excluded from the use or benefit of something; private property is created by the guarantee that an individual can exclude others from the use or benefit of something. (4–5)

In MacPherson's typology the crucial distinction is not between common property and private property, but between common rights and exclusive individual rights. The irony inherent in his typology is that common property is based not in common rights, but in individual rights guaranteed by the state. In fact, in MacPherson's categorization, common property is the only form of property that is invariably based in individual rights. At the other extreme, state property can never be based in individual rights but constitutes the corporate right of the state. Put another way, state property is neither common property nor private property:

State property, then, is to be classed as corporate property, which is exclusive property, and not as common property, which is non-exclusive property. State property is an exclusive right of an artificial person ... Common property is always a right of the natural individual person, whereas the other two kinds of property are not always so: private property may be a right of either a natural or an artificial person, and state property is always a right of an artificial person. (1978, 6)

In short, according to MacPherson, the state guarantees the right to common property and private property and may, as a corporate entity, hold property of its own. As a result, no individual has the right to exclude, as that right rests with the state, but certain individuals may be granted the right to use such property for their own exclusive benefit.

A major strength of MacPherson's analysis is the importance he places on the role of the state, for, if property is essentially not a thing but a right, then it is of primary importance to identify the source of

such rights. In identifying that source as the state, MacPherson is
taking part in a debate about the origin of the right to property that has
continued virtually unabated since at least the seventeenth century.
Most major social and political theorists have, in one way or another,
contributed to that debate.

Many of those theorists posited the existence of a 'state of nature'
before the development of individual property rights. For example,
the seventeenth-century political philosopher Thomas Hobbes ([1651]
1962, 132–41) held the view that individuals in a state of nature exercised their liberty to create a 'social contract' in order to avoid the
'warre of everyone against everyone else' that would otherwise result
from the clash of individual interests. Under such a contract they turned
over to the state (personified by the Leviathan) control over their individual rights. MacPherson (1962, 71) summarizes Hobbes's position as
follows:

Only by acknowledging such authority can men (a) hope to avoid the constant
danger of violent death and all the other evils which they would otherwise
necessarily bring upon themselves because of their otherwise necessarily de-
structive search for power over each other; and (b) hope to ensure the conditions
for the commodious living which they necessarily desire. Hence every man
who understands the requirements of men's nature and the necessary consequences of those requirements, must acknowledge obligation to a sovereign.

In the Hobbesian view, then, the power of the state is necessary to
enforce the rights of some individuals to property.

Another major British political philosopher of the seventeenth century,
John Locke, also posited a state of nature in which land was held in
common. However, he argued that individuals, by their own labour,
could move land out of a condition of common property. Labour invested
in property, and not the power of the state, was seen by Locke as the
source of the right to private property: 'we see in Commons, which
remain so by Compact, that 'tis the taking any part of what is common,
and removing it out of the state Nature leaves it in, which begins the
Property; without which the Common is of no use. And the taking of
this or that part, does not depend on the express consent of all the
Commoners ... The labour that was mine, removing them out of that
common state they were in, hath fixed my property in them' (Locke
[1689], quoted in MacPherson 1978, 18). Moreover, Locke ([1689] 1952,
17, 20) argued that individuals in a state of nature could collectively

contract to respect others' right to property without recourse to an external centralized power:

> God, who has given the world to men in common, has also given them reason to make use of it to the best advantage of life and convenience ... As much land as a man tills, plants improves, cultivates, and he can use the product of, so much is his property. He by his labour does, as it were, enclose it from the common. Nor will it invalidate his right to say everybody else has an equal title to it, and therefore he cannot appropriate, he cannot enclose, without the consent of all his fellow commoners – all mankind.

In a recent paper, Sabetti (1985) suggested that Hobbes's and Locke's approaches to the nature and origins of property rights are also the basis of two broader approaches to the nature of law. Sabetti distinguished between the 'Command Theory of the Rule of Law' and the 'Democratic Theory of the Rule of Law.' The Hobbesian view epitomizes for Sabetti the Command Theory, in which a single overarching structure of governmental arrangements is presumed to serve the public interests of all citizens. Such a view, he argues, 'allows little or no opportunity for citizens to act on a voluntary or coproductive basis' (12). By contrast, the Democratic Theory of the Rule of Law, which had its origins in the communal city-states of medieval Italy but which reached its fullest embodiment in the political analyses of Tocqueville and Jefferson, emphasizes the possibility and importance of the self-governing community.

Implications of the Theories of Property for the Regulation of Common Property

The preceding discussion has implications for our understanding of both the role of the state with regard to the regulation of property and the nature of property in all its forms. Although these two dimensions are interconnected, for analytic purposes they will be examined separately in this section.

Property and the Role of the State: The theories of the nature of property just outlined have important implications for our understanding of the 'tragedy of the commons' position. First, it is apparent that the conceptual origins of that position are located not in the works of Hardin and

Gordon, as is often implied, or even in the work of Lloyd (1933), to whom Hardin acknowledges an intellectual debt (see Hardin 1977, 6; McCay and Acheson 1987, 2–8), but rather in the theories of Thomas Hobbes. Moreover, as Sabetti's insightful analysis makes us aware, the 'tragedy of the commons' position, rooted in the Command Theory of the Rule of Law, rests on a particular set of assumptions about the nature of humanity, the role of the state, and the rule of law. When these assumptions are exposed, the implications of the 'tragedy of the commons' position become clearer.

The 'tragedy of the commons' position relies on a theory of human motivation that posits competition, if not outright war, as the natural condition of humanity. It follows from this that human cooperation to enforce the husbanding of property is impossible without a strong central state authority. Indeed, such an assumption suggests the necessity of a structure that puts the power to regulate resources entirely in the hands of the state. Those on whose behalf the state presumably acts have virtually no role in the regulation of the resource on which they depend. One recent work that does recognize the link between the Hobbesian position and the 'tragedy of the commons' thesis summarizes the relationship as follows: '[For Hobbes] the role of reason lay in agreements between individuals to submit to a common rule. The political state is a rational social contract. There can be no management of resources without political rule, a claim recognizable in the tragedy-of-the-commons thesis three hundred years later' (Stocks 1987, 109).

Our brief discussion of Locke's work, and Sabetti's description of the 'Democratic Theory of the Rule of Law,' suggest that there is an alternative to the Hobbesian/Command Theory position. That alternative is based on the assumption that cooperation is the natural condition of humanity, and therefore admits the possibility of cooperation with respect to the regulation of property as well. Indeed, it sees the self-governing community as an alternative to central state control in that regard (Matthews and Phyne 1988, 160–1). Sabetti extends his analysis to argue that the two historically different conceptions of the rule of law engender different ways of responding to community self-help movements. He argues that the Command Theory of the Rule of Law is antithetical to the rise of such cooperative movements because it gives virtually no credence to the possibility that they can be effective. By contrast, the Democratic Theory of the Rule of Law is conducive to a system of government that not only provides opportunities for cooperative, but actively works to promote or stimulate voluntary joint efforts

(Sabetti 1985, 26–8). Put somewhat differently, the idea inherent in the concept of common property is that of a covenant or moral consensus among users. A political or policy structure organized on principles consistent with Command Theory denies the very possibility of such consensus, whereas one organized around the principles of Democratic Theory seeks to engender it, and creates the conditions that make it possible. As things now stand, Canadian federal fisheries policy is organized around à set of assumptions that stem from principles consistent with Command Theory and, as such, denies the possibility of a social consensus among users to regulate fisheries common property in a way that enhances rather than depletes the resource.

The Nature of Property. There is some confusion in the literature with regard to the definition of property, especially as it pertains to distinctions among the different types of property. There appears to be a need, in particular, for further consideration of the distinction between common property and usufruct property and between state property and private property. Both advocates and opponents of the 'tragedy of the commons' thesis exhibit difficulty in delineating those distinctions.

Common property according to Welch's definition, cited earlier, is property that can be 'used by anyone.' He distinguishes it from usufruct property, over which users have the right to exclude non-users. This distinction, however, is conceptually inaccurate, at worst, and has little empirical utility, at best. Marchak (1987) has pointed out that, conceptually, the phrase *common property* is essentially a contradiction in terms. She argues that such a usage reflects a conception of property as a thing rather than a right: 'This use of the term implies that property is a thing, rather than a social arrangement of rights. As soon as we recognize the social source of property rights, this use of the term 'common' in association with 'property' becomes a contradiction, for if property necessarily involves a socially enforced set of exclusive rights, then a situation wherein there are no enforceable rights involves no property' (5). Marchak's position derives from MacPherson's definition of property as a right rather than a thing. But even if one chooses to view property as a thing, the point still holds that something over which one has no power to exclude others cannot easily be called one's 'property,' regardless whether 'one' is an individual or a collectivity (such as a community).

In addition to these conceptual problems, Welch's definition also presents empirical difficulties. As we shall see later in the chapter, there are few examples of common property anywhere in the world that would fit such a definition. In most situations where so-called common-property resources exists, there are myriad customs and regulations defining who has the right to use the resource, at what times, and under what conditions. At the very least, a distinction is almost invariably made between the access rights of local residents and those of outsiders and non-residents.

What appears to be required is acceptance of the possibility that 'common property,' as that term is normally used, is essentially an ideal type that rarely, if ever, occurs in reality. In reality what is usually referred to as common property is actually some form of usufruct property, wherein access to and use of the property is determined by one's membership in a 'community' of users. Membership in such a community is usually a prerequisite for acquiring the right to use the property, although it does not necessarily follow that all members of the community will be accorded that right. Indeed, such communities typically assume that they have the right to deny access to certain members of their community as well as to outsiders.

Once we recognize and accept that most, if not all, forms of what we call common property are in fact usufruct property, the next requirement is a typology of the usufruct property rights governing any resource. Such a typology should be able to distinguish the collective usufruct property rights that stem from community membership, such as those often found in the regulation of fisheries, from the usufruct property rights that stem from private ownership or state ownership, such as those granted to corporations engaged in such activities as oil exploration or forest harvesting. Such a typology should also incorporate a distinction between the *right to use* the property in question and the *right to exclude* others from doing so, as well as the criteria that govern exclusion. Furthermore, it should involve a consideration of the source of such rights and powers, examining whether it be the state or community membership or tradition. Finally, such a typology should take into consideration the interests in a resource that competing groups may have and the purposes for which they may want to control it (see Marchak 1987).

Although it is beyond the scope of our discussion here to develop such a schema, it is relevant to our purposes to examine in greater detail the factors that relate to the *source* of property rights. Inevitably, in any nation-state, all property rights exist within the context of the state and remain in force only with its sufferance; in other words, the state ultimately has the power to appropriate and regulate all property within its territory.[1] While this may be true, however, it is none the less open to question whether the state is the *source* of all property rights, as MacPherson and Marchak claim when they argue that the state 'creates the rights' to property in all cases. Such a position does not seem to take sufficient account of the many instances in which rights to property come from membership in a community or group within the state rather than from membership in the state itself. The possibility that there are sources of property rights other than the Leviathan, as Hobbes would have us believe, or the state, as MacPherson has argued, must be given further consideration. For example, 'squatters' rights' grant possession of private property to squatters even when such possession is contrary to the more formal law of the state. Similarly, community customs, and norms and traditional practices are the source of certain property rights enjoyed by the residents of a community.

Such property rights form the basis of the claim by fishing communities that they have traditionally controlled the common property on which their survival is based. If such rights to use property and to exclude others from its use exist, they must have originated in a society based on principles consistent with the Democratic Theory of the Rule of Law rather than the Command Theory of the Rule of Law.

This analysis leads us to reformulate our conception of common property. At least in the context of the nation-state, the legitimacy of common property does not rest on its openness and availability to all (a condition, as we have already pointed out, that precludes the notion of property as a right in any case). Rather, the right of common property rests on the acceptance by the state and others that the claims of a local community to regulate in the interests of its members as a whole are legitimate. As part of its right to regulate common property, the community is invested, by tradition and custom, with the right to *exclude* others. (It should be noted here that any situation in which an individual has the right to 'sell' his or her property rights must be considered something other than a common-property situation. Transferable rights can involve only private, not common, property.)

It might be argued that, although the analytic distinction between community rights and state rights implied in the preceding discussion might have had some legitimacy in the past, it is not a relevant distinction today because the state has simply replaced the community in its regulating capacity. Marchak has cogently argued however, that it is difficult to sustain such a claim in view of the substantial difference in kind between a community's property rights and those of a state. This difference is rooted in the different interests that communities and governments serve and the differential power that those interests have to influence the actions of the state. As Marchak (1987, 11) puts it,

the community sharing a commons had equal interests in that area, and all members needed the resources in both the short and long terms. Further, all members could, physically, impose sanctions on others because such communities were small ... The participants in the commons were roughly equal in power, and no one party could impose restrictions on others.

Governments, by contrast, are institutionally constructed so as to manage not one resource but many; and they are required to balance, negotiate, and make decisions about conflicting interests. Not all members of the population have an immediate interest in any resource, and the interests that exist have divergent and conflicting requirements ... Some participants are vastly more powerful than others, and the actions of the more powerful can and often do preclude action by others. And all of these conditions rest on a prior fact: the resources are no longer used primarily for subsistence. They are now potential commodities.

In the context of our earlier discussion of the 'tragedy of the commons' position, the foundations of the competing interpretations of the role of the state in the regulation of the fishery become clearer. If the 'tragedy of the commons' perspective is rooted in the Command Theory of the Rule of Law, then the alternative is based on a theory of common property that stems from the Democratic Theory of the Rule of Law. The latter defines common property not as being open to all to use equally, but simply as *community property*. Common property is community property. The right to both regulate its use and exclude others from access to it is seen as stemming from the community. The 'tragedy of the commons' perspective simply ignores community rights and contends that all common property is essentially state property.

As noted earlier, the other dimension that must be considered in any classification of types of property is the relationship between state property and private property. As MacPherson clearly defines it, state property is the exclusive preserve of the state. It is true that the state may grant the right to use such property to certain individuals or groups, but such usufruct rights do not alter the fact that ownership of the property remains with the state. By contrast, in the case of private property, individuals or groups can claim the right both to use the property in question and to exclude others from its use. In addition, they have the right to sell the property. It is this right to dispose of ownership that appears to be the distinguishing characteristic of private property.

In this context, the process whereby the state awards fishing licences, as it does in Canada, starts to take on new meaning. At issue is the status of the licences as property. If they cannot be sold or traded by the individuals who possess them, they cannot be defined as private property. Rather, they are possessed under usufruct rights granted by the state to some persons but not others. If, however, the state grants a licence to an individual who then has the exclusive right to sell it to another person, that licence would appear to be private property. To that extent, the possession of fishing licences on Canada's East and West coasts can be said to involve very different types of property relationships. On the East Coast, individuals do not have the right to sell their licences, so the licences remain, for all intents and purposes, state property; on the West Coast, individual licence holders do have that right, so their licences can be considered private property.

The nature of a fishing licence as a form of property should not be confused with the nature of the fish as a property resource. Whereas the licence to fish may be either private or state property, the fish themselves would appear to be either state property or community common property. They are state property if one accepts the assumptions of the 'tragedy of the commons' perspective. They can be considered common property if one accepts the earlier argument that common property derives its legitimacy not from being open to all, but rather from being the traditional preserve of a community of users. Perhaps the best way to summarize the position taken here is to state that common property is essentially social property – the preserve of a community of users who have traditionally maintained the right to regulate both its use and the exclusion of others from it. Community is the social basis of common property. Common property is, in effect, community common property.

A review of terminology may be in order before we proceed. On the one hand, we have argued, with Marchak, that the term *common property*, as it is normally used, is a logical impossibility, as it precludes the existence of property rights of any sort, and, therefore, that it cannot exist in reality. On the other hand, we have attempted to demonstrate that what is generally empirically observed and termed 'common property' is, in fact, a form of 'community common property,' in that the usufruct rights derive not from the state or from private ownership, but from membership in the community of users. We are now faced with the problem of deciding what term to use when describing so-called common property. Because the term *common property* is by now embedded in the literature on the subject, we will continue to use it, with the understanding that we are referring to the type of community usufruct property rights we have described here. Wherever possible, however, we will use the phrase that more accurately reflects our understanding of the phenomenon – *community common property*.

Commons and Community

Opponents of the 'tragedy of the commons' position have generally recognized, either explicitly or implicitly, the link between community rights and what is called common property. For example, the foreword to the published proceedings of a 1985 conference on common-property resource management organized by a branch of the National Research Council in the United States argues that

when practised at the community level, this common property resource management has often been successful, or the resources would have vanished long since.

Restoring to the community the responsibility that was originally its own may be our only hope for the future protection of our soil, water, fisheries, pastures, forests, and wildlife. (Swaminathan 1986, v–vi)

Taylor (1987) has even argued that the problem with the 'tragedy of the commons' model is that it is based on the assumptions about human motivation derived from classical economics – assumptions that give little credence to the possibility of community property rights and management. In his words,

Following Hardin (1968), the theoretical literature on common-property systems has so far been dominated by the classical economic model of self-seeking and essentially unconnected individuals. Communities, from that perspective, have only a hypothetical and largely irrelevant existence. Indeed, they may be shown to be impossible *per se*. Many anthropologists find such a model inadequate, given the fact that they often find themselves in actual communities possessed of some common resource more or less successfully managed. (291)

As this quotation indicates, much of the opposition to the 'tragedy of the commons' perspective of the past decade has come from those who have been actively involved in research on small, resource-based communities. Numerous studies from around the world have explored the common-property character of resource exploitation in fishing and farming communities. Although it is not possible to summarize all of these studies here, we will consider two bodies of work, by James Acheson and by Anthony Davis and his colleagues, that deal with fishing communities on the East Coast of Canada and the United States. They are of particular relevance to this study because they describe community-based fishing organizations quite similar to those in Newfoundland and because they contain conceptual arguments about the nature of common-property relations that provide us with further insight into the practice of common-property management in such communities.

Acheson's (1975) study of lobster fishermen in coastal Maine was one of the first empirical studies of community-property relationships among fishermen, and it has become a frequently cited classic in the common-property literature. (An expanded version of this work appeared in 1987.) Acheson (1975, 187) points out that, 'from the legal view, anyone who has a licence can go lobster fishing anywhere,' but that, in fact, local fishing communities in Maine have developed their own 'property' norms that dictate who can fish for lobster, and where:

To go lobster fishing at all, one needs to be accepted by the men fishing out of one harbor; and once one has gained admission to a 'harbor gang', one is ordinarily allowed to go fishing only in the traditional territory of that harbor. Interlopers are met with strong sanctions, sometimes merely verbal, and more often involving the destruction of lobstering gear. This entire territorial system is entirely the result of political competition between groups of lobstermen. It contains no 'legal' or jural elements.

Acheson describes two types of property arrangements, which he labels 'nucleated' and 'perimeter-defended.' The nucleated form of ownership appears to be community based, in that 'men from each harbor gang have a strong sense of territoriality close to the mouth of the harbor where they anchor their boats' (1975, 189). Perimeter-defended ownership, by contrast, appears to depend not on membership in a community but rather on ownership of land adjacent to the fishing grounds. Acheson suggests that this system of perimeter-defended ownership was once widespread (1987, 45), but is now used primarily as a way of controlling right of access to fishing grounds near offshore islands: 'In both types of territories, but most clearly in the perimeter-defended areas, claims over ocean areas are tied up with formal ownership of land. Ownership of land on an island is held to mean ownership of "fishing-rights" in nearby waters, despite the fact that legally the ocean areas are part of the public domain. On Matinicus, for example, no one is allowed to fish in the island's territory unless he owns land on the island' (1975, 190). Even more interesting is Acheson's observation that such rights can be rented or sold:

A major argument against selling land to 'summer people' is that thereby an island family may lose its fishing rights ... In perimeter-defended areas, owner-ship rights to the waters are not merely usufructuary. Even if the owner is not using his water territory, his fishing rights remain, and may be rented out. In these areas, ownership rights are so strong that men who own whole islands rent out 'water areas' to men from nearby mainland harbours ... Arrangements vary considerably, but in some cases the families take half the gross income of these renters as return on capital equipment and as rent on the 'water area'. These rent rights are traditionally held, and inherited patrilineally, as are land property rights. (1975, 190–1)

Acheson establishes that, in at least one inshore fishing area, local residents treat fishing grounds as 'property' even when there is abso-lutely no formal legal recognition of such property rights. However, he appears to have confused the nature of the property relations that are involved. Thus, in a section of his recent work entitled 'Controlling the Commons,' he implies that both the nucleated and perimeter-defended forms of property rights are common-property rights. Indeed, he 'deplore[s] the breakdown of the perimeter-defended areas ... given what we know about the theory of common property resources' (1987, 63).

Acheson seems to have confused two issues. The first is whether the property rights in question derive their legitimacy from the state or the community. The second is whether the rights in question are common property, private property, usufruct property, or state property. There can be no doubt that, if Acheson's description is correct, neither nucle-ated nor perimeter-defended property rights derive their legitimacy from the state. Rather, the legitimacy of both is based on accepted norms of the community. It does not follow from this that both nucle-ated and perimeter-defended property rights are forms of common prop-erty.[2] Acheson's own description of perimeter-defended property rights suggests that they respect a form of private property rather than common property. Individuals can rent or sell them, and when they do, they are not required to seek communal approval or to turn the profits over to the community. It would be hard to find a clearer example of private property – albeit private property for which the legitimacy of ownership is derived from community custom rather than from the state. In contrast, Acheson's description of the property relationships he refers to as 'nucleated' puts them within the definition of usufruct common property or community common property, as that concept was outlined earlier. Access and usufruct rights governing nucleated property are controlled by the community, and individuals have no recognized right to buy or sell access rights for their own benefit, or even for the benefit of the community. In sum, Acheson's most important contribution might reside in his documentation of the fact that both private property rights and usufruct property rights may derive their legitimacy from tradition rather than the state.

In a series of recent papers (Davis 1984, Davis and Kasdan 1985, Davis and Thiessen 1986), Anthony Davis and his colleagues have examined the nature of community-based property rights among fishermen in southwest Nova Scotia. Davis describes the property rights that have been developed by lobster and longliner fishermen in three communities that fish in an area he calls Port Lameron Harbour. He contends that the fishermen base their claims to these rights on three grounds: their continuous occupation and use of the fishing grounds, the proximity of their fishing ports to those fishing grounds, and their economic dependence on the resource (1984, 145). Davis makes explicitly clear that such rights are not private property rights but usufruct rights that derive their legitimacy from community membership:

The individual fisherman's right to exploit the resource zone is based on a form of usufruct or use-right ... Claims of ownership and control of property is [sic] centred in the community, and individual use-rights are derived from membership in the community. Although basically foreign to the system of the owner/non-owner property relations dominant in capitalist-industrial societies, collectively based property claims and associated individual use-rights represent a distinctive form of property relation. (145–6)

However, he also argues that these property rights are not rights to common property in the usual sense of that term, which, he contends, essentially denies the existence of property relations at the community level. He does maintain, however, that the informal and formal systems of community-based property relations that he describes constitute a clear set of property relations that specify and regulate the conditions of access to and use of the resource:

The common property view assumes that, since marine resources are not directly and legally owned by individuals or institutions (in a manner analogous to land-based resources), access and use is open to all. But a condition of open access would not lend itself to satisfying the imperatives of people dependent upon exploiting marine resources for their livelihood. Informal and formal management rules and systems constitute property relations insofar as they specify, protect and reproduce the conditions of access, i.e., they limit and legitimate access and use of marine resources. (140–1)

As we argued earlier in the chapter, common property, as that term is usually employed, is not property at all. While it admits the possibility that communal property rights can derive from a source other than the state, that concept does not recognize the right of users to limit the access of others. Davis's work supports our interpretation, demonstrating that, although property relations abound in Port Lameron Harbour, they are neither common-property relations (in the usual sense of that term) nor private-property relations. Rather, they are usufruct property relations: individuals have personal usufruct rights, deriving from membership in the community, to use the resource and to exclude others from its use, but they cannot buy or sell those rights – that is, they have no private-property rights to the resource. The reader will note that this system is quite different from the private-property relations that Acheson described as existing in the perimeter-defended areas of the Maine lobster-fishing grounds.

There have been no similar studies explicitly examining the nature of property relations in the Newfoundland inshore fishery, although the subject has been addressed in passing in several studies with a somewhat different focus. For example, Martin's analysis of the inshore fishery in Fermeuse (1973, 1979) describes two forms of local regulation. In the inshore cod fishery the trap 'berths' used by local residents were closely regulated. The same boat crew could apparently use a particular berth every year, without being challenged by other crews. Also, as more-modern and more-mobile fishing technology was introduced, certain fishing areas were set aside for the exclusive use of fishermen who lacked the new equipment and still used the traditional technologies of handlines and traps. McCay (1978) also touches on the

TABLE 4.1
Taxonomy of the allocation of fishery property rights

| | | Source of legitimation of property rights | |
		State	Community
Type of property right granted	Usufruct	non-marketable licence/quota	nucleated
	Private	marketable/ transferable licences and quotas	perimeter-defended

issue of property rights in Newfoundland in her discussion of the introduction of longliners in the community of Fogo. She describes a situation in which the introduction of a new 'class' of longliner fishermen created conflict over rights of access. In a more recent work McCay has argued that, from their origins, common-property rights have been 'grounded in the class struggle' (1987, 202), presumably between those who benefit from usufruct rights to community common property and those who attempt to appropriate common property for their personal use. She suggests that these inherent conflicts are reflected today in the groups using different technologies of production (1987, 205).

Our analysis of existing studies suggests two dimensions in terms of which the allocation of property rights may be analysed: (1) the source of legitimation of such allocation, and (2) the type of property right involved. These, in turn, can be cross-referenced to create the taxonomy depicted in Table 4.1. This framework schematizes the relationship between property rights legitimized by the state and those legitimized by community values, norms, and practices. However, it does not explain why community common-property relations persist in the face of state regulation. That question can now be addressed in the context of the attributes of community common property.

The Attributes of Community Common Property

Why do community common-property relations persist, at least in small resource-based communities, if there is as great a potential for them to be undermined as the 'tragedy of the commons' argument would have us believe? The answer would appear to be that, for certain types of resources and under certain conditions, such systems represent a 'functional' way of managing property and ensuring that the greatest number of individuals or households possible have the opportunity to earn an adequate living. They have proved most functional in situations in which there is no guarantee that the 'harvest' will remain relatively constant in any given location over an extended period of time (see Netting 1976, 140; Ostrom 1987a, 256). Under such circumstances the private ownership of a single location may not be as profitable in the long term as the right to use a variety of different locations. Furthermore, even if the ownership of a single location were more profitable than the ability to use several locations, it may not be considered preferable in the long term because it does not have the insurance value that is inherent in the right to use a variety of locations. Indeed, the long-term functionality of common property may have less to do with maximizing gains and profits than with providing insurance against 'crop failure' in the future.[3] In addition, communal ownership is advantageous in any situation in which it is necessary to harvest a variety of different resources found in different ecological habitats. Finally, even when the harvest is predictable and secure, communal ownership may provide the opportunity for different participants to benefit over a period of time, thereby equalizing both opportunity and results. It follows that communal common-property relationships are more likely to be found where the participants are relatively poor.

The aforementioned attributes are all prominent characteristics of the inshore fishery in Atlantic Canada. Since many species do not necessarily appear in the same place annually, a system of communal common property may allow fishermen greater geographic mobility than would otherwise be possible. Furthermore, most inshore fishermen find it necessary to engage in a diversified fishery, seeking many different species that are often found in quite widely dispersed habitats. Under such circumstances the ownership of a single fishery location would likely prove impractical. Finally, the system of access regulations in some communities ensures that purely chance rewards will be more equitably distributed, thereby contributing to equality of opportunity and results.

In such situations communal common property is not necessarily the anachronism that some advocates of the 'tragedy of the commons' theory would have us believe. Rather, it is an economically and socially 'rational' solution to a distinctive set of circumstances. Nevertheless, the problem that is basic to common-property situations is the one on which the 'tragedy of the commons' argument focuses – namely, how access to and use of common property can be regulated to prevent overuse and depletion of the resources in question. As Oakerson (1986, 13) has noted, 'All common property nevertheless faces one common problem: how to coordinate individual users to attain an optimal rate of production and consumption for the whole community.'

One body of analysis that touches on the subject of the regulation of common property is known as 'the new institutional economics' (Oberschall and Leifer 1986). Works in this area deal with the nature of efficiency in non-market institutions. The new institutional economics takes the position that institutions arise and persist when the benefits they give are greater than the 'transaction costs' (in the case of the common-property situation, the social and economic costs) involved in creating and maintaining them (Oberschall and Leifer 1986, 237). Viewed from this perspective, common-property relationships are understandably most prominent where users are relatively poor. In such circumstances the access provisions associated with common property tend to help equalize rewards. By contrast, 'prosperity makes sharing groups smaller and more numerous until most goods become completely privatized in single-member groups' (Oberschall and Leifer 1986, 244). However, such a perspective tends to ignore or downplay the power of external agencies (such as the state) to influence and undermine such cooperative arrangements, even when those arrangements remain functional for the participants. Yet, as was shown in chapter 3, that is precisely what has occurred in Canada's East Coast fishery.

Another body of analysis that treats issues involving the regulation of common property is found in the work of political scientists engaged in 'public-choice theory.' Public-choice theorists focus on the factors that impede or enhance the use of common-pool resources. Unlike the new institutional economics, public-choice theory tends to take explicit account of the role of the state in either protecting or undermining existing community common-property arrangements.

Of particular interest here is the works of Ostrom and her colleagues (Blomquist and Ostrom 1985; Ostrom 1987a, 1987b, 1988), which focuses explicitly on the ways in which individuals in common-property situations may 'extricate themselves from Commons Dilemmas by establishing self-governing institutions' (Ostrom 1988, Lecture 1, p. 16). Blomquist and Ostrom focus on the institutional arrangements that enhance the capability of individual participants to resolve the dilemma of the commons. They emphasize the importance of establishing (1) institutionalized channels of communication and information among users, (2) institutional arrangements for enforcement of contracts, and (3) effective monitoring of use-patterns (1985, 389). More recently, Ostrom (1988) has reviewed studies of successful common-property systems throughout the world, and developed a list of eight characteristics that she contends are basic to them (Lecture 2, pp. 41–2). Successful common-property institutional arrangements are said to have:

1 defined a limited set of individuals who are eligible to obtain resource units from a commonly owned resource system;
2 carefully specified the rights and duties of membership;
3 allocated resource units roughly proportional to the contributions that are required;
4 developed monitoring arrangements that reduce the opportunities for individual appropriators to take more resource units or to shirk in contributing inputs without detection;
5 created a governance system which gives most members an active voice in determining the rules to be used;
6 relied on members of the resource community to serve as officials and guards;
7 developed conflict-resolution mechanisms that are prompt and involve low transaction costs;
8 developed regularized procedures for changing the rules in use under different environmental conditions.

Whereas Blomquist and Ostrom's earlier list focused primarily on the importance of channels of information and communication, virtually all of these eight items deal directly or indirectly with the nature of regulation, specifically with the development or enforcement of rules.

Ostrom extends this analysis of rules to address the relationship between common-property regulation and state regulation more directly. Based on her 1988 analysis of studies of successful common-property systems, she argues that, if common-property relationships are to continue, the customary rules developed by local communities to police their resource must be 'nested' within the larger set of state rules (Lecture 2, p. 44). In other words, state rules should, at the very least, not explicitly interfere with customary regulation of common-property resources; indeed, if possible, they should explicitly recognize the legitimacy of such local regulatory practices. It is in this context that Ostrom examines the regulation of Canada's East Coast fishery as described by Davis (1984) and Matthews (1988). She contends that, unlike the successful common-property systems on which her analysis is primarily based, the Canadian Atlantic fishery is best described as an example of a 'fragile' common-property resource institution (Lecture 4, p. 29). She concludes:

If [Canadian] national policies were to change and to try to develop a set of nested rules that helped enforce local regulations developed over the years while focusing most of the new regulatory effort on the far offshore fisheries which are indeed open access, this 'fragile' rule system could survive, adapt, and enable fishermen in future generations to make effective use of this local resource indefinitely into the future. However, if Canadian policy continues to try to develop a single, coherent policy for all fisheries along the entire East Coast, then I fear the eventual deterioration of the locally evolved system. Further, I doubt that any national agency has the extensive time-and-place information needed to tailor a set of rules to the particulars of local situations. (36).

As this discussion illustrates, the focus of Ostrom's attention is essentially on the normative aspects of institutional arrangements. Her primary concern is to determine the social structural norms (that is, rules) that enable common-property relations to continue. From this perspective, the factors that motivate individuals to participate in common-property usufruct relations are seen largely in normative terms. Indeed, when describing her eight attributes of common-property resource systems, Ostrom states, 'The first seven attributes enable these systems to motivate participants to contribute effort to achieve collective benefits' (Lecture 2, p. 42).

There is, however, an alternative perspective on human motivation that does not view it as just compliance with norms and institutional structures. It argues instead that people are motivated through 'constructed action'; that is, they reconstruct their social reality in their own minds, then act on that construct (see Turner 1962; Berger and Luckman 1967; Schutz 1971a, 1971b). It will be argued in the following section that this alternative perspective on human motivation and behaviour provides a useful framework for analysing the nature of individuals in common-property situations such as the ones that we describe later in the book. This perspective complements the work of

Ostrom and other public-choice theorists and provides a consideration of dimensions of human action and behaviour that have not been adequately addressed in the existing literature.

A Social-Psychological Perspective on Common-Property Relations

One reason much of the existing literature focuses on the normative rules underlying common-property relations is that its primary consideration is the nature of the regulatory and legal arrangements on which common property is based. Regulations and laws, after all, are the embodiment of cultural values and norms. Indeed, to determine the fundamental values and norms of any society, an obvious place to start is with the values inherent in that society's laws.

Another reason for this focus on norms is the orientation of most of those who have dealt with this subject. Virtually all have approached property relations from a structural perspective; that is, their concern is with the structural factors that contribute to the nature and maintenance of property structures, and not with the nature of the human interaction that occurs within them. Although they may note that property relations can take on a variety of meanings for those involved in them, they pay little attention to the factors that influence such meanings or to the way such meanings become determinants of human interaction.[4]

Still, the social-psychological aspects of such property arrangements have not been ignored entirely. Indeed, as was argued earlier, the 'tragedy of the commons' perspective is premised on assumptions about the nature of human motivation under conditions of open access and common property. Similarly, most of the works that have attempted to provide alternatives to the 'tragedy of the commons' perspective have touched on dimensions of property relations that require an understanding of the social-psychological interpretations and meanings underlying human interaction. For example, the quotation from C.B. MacPherson's work on property (1978) at the beginning of this chapter is essentially a social-psychological statement. To argue, as he does there, that property relations are not constant but rather change over time in relation to the 'meanings' that people give them is, essentially, to take a phenomenological position. In other words, MacPherson's statement is consistent with the argument that man lives and operates in a socially constructed world in which the nature and 'meaning' of institutional relationships are matters of individual interpretation and definition, subject to change over time. This position is also related to Gidden's (1979, 1984) argument, as outlined in chapter 1, that individuals are not simply 'cultural dopes' responding to social-structural forces, but active agents attempting to make sense of and manipulate the world around them. It is implicit in both arguments that individuals give meaning to the world around them, then act on the basis of that meaning.

The place of social-psychological factors in common-property relations is also readily apparent in Oakerson's attempt to develop 'a model for the analysis of common property problems' (1986, 13–29). His model focuses on four components: (1) technical and physical constraints related to the physical and social environment in which common-property relations operate; (2) decision-making arrangements, particularly as they pertain to the development of rules; (3) patterns of interaction; and (4) the nature of the outcome of such relationships. In this schema, social-psychological considerations are associated primarily with the third component, patterns of interaction. Oakerson (1986, 21) states:

Given the technical and/or physical features of a commons and the decision-making arrangements available to govern it, the next question concerns behaviour: what patterns of interaction characterize the behaviour of users and other decision makers in relation to the commons? ... What matters is how individuals choose to behave in relation to one another. Patterns of interaction derive from mutual choice of strategies; that is, each individual's choice of strategy (how to relate to others) depends upon individual expectations of others' behaviour.

However, although Oakerson clearly identifies the need to focus on the social-psychological factors of individual choice in the analysis of common-property relationships, his 'theory' of social psychology is limited largely to the 'costs and benefits' approach of economics and formal game theory. In his words, 'choices are generally viewed in terms of a comparison of costs and benefits' (1986, 20). In contrast, it is the argument here that a comprehensive knowledge of the social psychology of common-property relationships requires a broader understanding of social-psychological processes and human interaction than is represented in the cost–benefit perspective.[5]

One starting point for gaining that broader understanding is an in-depth exploration of the nature of *commitment*. Commitment, particularly *commitment to community*, can be seen as the foundation of common-property usufruct relationships. It is commitment to community that prevents behaviour motivated by self-interest, as reflected in the cost–benefit approach, from dominating and, in doing so, averts the tragedy of the commons. Fikret Berkes, in a series of studies of various types of fishermen, including native fishermen on James Bay (1977, 1987), commercial fishermen on Lake Erie (1983, 1985) commercial and subsistence fishermen in British Columbia (1983) and inshore fishermen in Turkey (1983; 1986), has emphasized the importance of commitment to community. All of his studies establish that 'in a traditional fishing society, it is difficult if not impossible to be a selfish user who can ignore community concerns' (1983, 14). Similarly, Ostrom (1987a) takes exception to game theory and social-choice theory for their acceptance of 'entirely opportunistic behaviour as the universal norm of behaviour' and their failure to realize that 'a theory of totally opportunistic behaviour will not predict the use of institutions that rely to a large extent on self-enforcement or mutual-enforcement of rules rather than on the active presence of external policies.'

In the common-property literature, little attention has been given to the nature and dimensions of commitment to community. In another context, this writer has argued that two prerequisites of commitment to a group, a location, or an ideal are (1) identification of that group, location, or ideal as important, and (2) identification with it on a personal level (Matthews 1983, 22–4). Such social-psychological processes would appear to be the necessary precursors to the development of community common-property relationships, as well as being the necessary social-psychological preconditions for them to survive. What is needed, then, in addition to a focus on the normative base of regulations and rules, is an analysis of the dimensions of commitment and identification that form the basis of communal common-property relationships.

The consideration of such factors in the analysis of common property highlights the fact that such property relationships retain a symbolic and somewhat metaphoric character in their own right. A community is essentially a 'shared universe of meanings' in which members recognize themselves and their relationships with others. Gary Alan Fine (1979, 1983) highlights the implications of these shared meanings for group solidarity, as evidenced in the following passage (1983, 13):

Every group develops a culture which I have termed its idioculture. An idioculture is a system of knowledge, beliefs, behaviours, and customs peculiar to an interacting group to which members refer and employ as the basis of future interaction. Members recognize that they share experiences and that these experiences can be referred to with the expectation that they will be understood by other members, and can be employed to construct a shared universe of discourse.

In communities in which they persist, community common-property relationships have the potential to constitute a substantial foundation for their members' 'shared universe of meanings.' Indeed, they may constitute much of the symbolic structure that distinguishes a community from others in the minds of its members. They also represent a significant part of the 'metaphor of reciprocity' that binds the community together (Fernandez 1987, 284). Indeed, the symbolic character of common-property relationships forms at least part of the basis on which members establish identification with and commitment to their community.

'Because Fish Swim': The Interrelationship of Property and Territoriality in the Regulation of a Common-Property Resource

Virtually by definition, property relations involve the specification of territoriality and the control of resources. This is particularly true in the case of community common property. Yet the attributes of the relationship among property relations, territoriality and resources have not been clearly delineated. Before concluding this chapter, it seems appropriate to give this issue some consideration.

There is a wonderful story by Newfoundland author Ted Russell (published in his *Chronicles of Uncle Mose* [St. John's: Breakwater Books, 1975]) about a dispute between two Newfoundland fishermen as they attempt to catch fish during the winter through holes cut in the ice of a frozen harbour. One of the fishermen accuses the other of 'stealing his hole.' In due course, the person so charged finds himself before the local magistrate. In his defence, he does not deny that he took the other fisherman's hole, but argues instead that the hole cannot be regarded as property since it was cut out of something and occupied by nothing. Holes, he implied, are simply lines around nothing and, by definition, cannot be regarded as property.

While the fictional Newfoundland fisherman may be correct in arguing that it is possible to have space (or territory) without having property, it is much more difficult to have property without space. This is particularly the case when property is a resource, for a resource is usually fixed in space. Fish, however, are an exception, as they are able to swim in and out of a particular space. Thus, if location is the defining characteristic of property, then at any one given time, depending on their location, fish may or may not be the property of a particular person or community. The basis of the claim by the Canadian government that it has the exclusive right to regulate access to the East Coast fishery rests, to a considerable degree, on the argument that it is the only jurisdiction big enough to emcompass the whole 'territory' in which fish swim. In this formulation, the basis of the state's claim to property rights over the fishery can be expressed quite simply: 'As then Canadian Prime Minister Pierre Trudeau observed, in reference to the need for a Law of the Sea, it was essential "because fish swim" (Marchak 1988, 10).

As Marchak puts it, 'The argument is that the mobility of fish is the reason for their "common property" status, and their property status, in turn, is the cause of resource depletion' (1988, 20). Indeed, the justification frequently given by the federal government for assuming control over the regulation of the fishery is that, because fish swim, 'fish in water' are nobody's property and must therefore be regarded as state property. 'Fish in the boat,' however, can be regarded as property; they become the private property of those who have harvested them with state consent. In such a formulation, there is no place for any consideration of fish as common or communal property. It is also argued that, again, because fish are mobile, the state must step in to protect the interests of those whose territory the fish have vacated, at any given time, from potential overfishing by those whose territory they have occupied at that time. Thus, the statement 'because fish swim' becomes a declaration that resource property rights are in some way related to location and that, unless it can be shown that the resource consistently occupies a specific location, the only sustainable property rights are state property rights.

This formulation would also appear to involve an analytic distinction between two *types* of property. One is the local territory through which fish swim, and the other is the resource itself – the fish. From this perspective, disputes between fishermen and the state about the right to regulate the fishery can be seen to arise out of very different conceptions of the relationship between territoriality and property rights. Fishermen, it would seem, regard local territorial proximity as the defining characteristic giving them the 'right of property' and, with it, the right to regulate access to the fishery as well as the right to catch fish in their local territory. In contrast, government regards local territorial proximity as the very reason that fishermen cannot have the 'right of property' when it comes to the regulation of the fishery. In sum, the government focuses its 'right of property' claim on the resource itself (the fish) wherever that resource is located, and argues that control over the resource takes precedence over territorial proximity. Even when the government does consider territoriality, it regards not the proximity of the territory to the resource but the size of the territory as the characteristic that defines the property relationship.

In one of the few attempts to produce a theoretical framework for the analysis of territoriality, Lyman and Scott (1970) distinguished four types of territories: public, home, interactional, and 'body.' While the latter two have little relevance here, the distinction between public and home territories is useful in understanding the different symbolic meanings that attach to territory when property is being defined. Public territories, according to Lyman and Scott, 'are those areas where the individual has freedom of access, but not necessarily of action, by virtue of his claim to citizenship.' Although such territories are 'officially open to all,' Lyman and Scott argue that, in reality, 'certain categories of persons are accorded only limited access to and restricted activity in public places' (91). Lyman's and Scott's depiction of public territories is consistent with the federal government's definition of state property. State property rights largely involve the control over and limitation of access to public territory.

In contrast, Lyman and Scott define home territories as 'areas where the regular participants have a relative freedom of behaviour and a sense of intimacy and control over the area.' That 'sense of intimacy' is said to be related to the 'territorial stakes or identity pegs' that habitual users have in the area. Lyman and Scott note, however, that 'home and public territories may be easily confused,' in that home territorial status may be lost through violation of established usages or as a result of conquest, whereas 'public areas, precisely because of their open condition, are vulnerable to being converted into home territories' (1970, 92–4). Lyman's and Scott's description of what they call home property essentially captures the territorial basis of the community common-property relations we have described here. Especially significant is the emphasis they place on the extent to which the 'sense of intimacy and control' over an area derives from habitual users' 'territorial stakes or identity pegs' in it. In the inshore fishery, local fishermen stake their property claims on the fact that local fishing grounds are their 'home territory.' Such claims are intimately related to the fishermen's own sense of identity and to their identification *with* their community. The state does not recognize such claims because it maintains that the mobile nature of the resource renders territorial considerations invalid.

Conclusion

This chapter has focused on the broader theoretical and conceptual context of fisheries regulation. It has attempted to shed light on the nature of property relations particularly as they pertain to common property. It has argued that common property as defined in the 'tragedy of the commons' formulation is not property at all and, in all probability, has existed only in theory. This implies that the current regulation of the East Coast fishery is founded on assumptions that have little basis in reality.

According to this chapter's analysis, the defining attributes of so-called common property are (1) that usufruct rights to property are seen to derive from territorial proximity to a resource, and (2) that those rights are held collectively rather than privately. Support for this was found in a variety of empirical case studies.

However, the link between commons and community is not only structural, but also symbolic. We have shown that the existing literature on common-property relations has limited itself primarily to structural factors and rules, without giving much consideration to the symbolic and social-psychological aspects of property relations. As MacPherson noted, the meaning of property relations is not constant but changes over time; it is clear that scholarly attention needs to be directed to the social-psychological dimensions of such reconstructions of meaning. That analysis should focus on the identification with and commitment to community, and on the way in which those factors are influenced by the nature of the property relationships involved.

The chapter ended with a look at the interrelationships among property relations, territoriality, and resources. We have suggested that differences between the government's and fishermen's claims to the right to regulate the inshore fishery may lie in differing conceptualizations of these interrelationships.

5 Small Worlds

The Calculus of Work and Survival in
Two Small Fishing Communities

We're not greedy here. What's put in this world is put here for everybody.
And for one or two to want it all, I don't see it.

A Charleston fisherman
(Respondent no. 003)

The way it is with us here, we got no problems. We got no trouble getting
clear of our fish and getting a good price for them too.

A King's Cove fisherman
(Respondent no. 016)

One of the underlying themes of this book is the diversity of fishing
communities. This chapter will examine two of the smaller fishing
communities in Newfoundland, both with fewer than 300 residents. Yet,
in the context of Newfoundland, such communities cannot be deemed
insignificant. As late as 1971, 71.4 per cent of all Newfoundland communities had fewer than 400 persons, and 62 per cent (some 545 communities) had fewer than 300 persons (Matthews 1976, 19). To be sure,
a considerably larger proportion of Newfoundland's population is living
in the remaining 28.6 per cent of larger communities. However, as the
small communities are the ones most likely to retain elements of traditional fishing practices, they provide us with a unique opportunity to
observe the old ways in the context of new demands. In other words,
an examination of the social organization and behaviour of fishermen
in these small communities enables us to observe how traditional practices are being altered by current events.

Charleston and King's Cove are the two communities described in this chapter. As we shall see, they are in many ways strikingly different. They look different. They engage in somewhat different types of fisheries. Indeed, the organization of fishery work in the two communities is different. It is these differences that are the focus of the chapter. None the less, the two communities also have much in common. In both, each of the people we interviewed were engaged in a series of personal calculations – their own 'calculus' of how best to make an adequate living in the physical and social environment in which they find themselves.

The process of making a living in this environment is constrained in a number of ways. First, it is constrained by the physical environment itself. Those of us who live in more urban environments are apt to be relatively unaware of the impact of our physical surroundings on our lives. In the small resource communities of Newfoundland, such inattention or ignorance would be unthinkable. Here, the environment sets the conditions for survival itself. Proximity to the fishing grounds, the capricious migratory behaviour of the fish, the availability of alternative resources that might provide employment, and the weather itself are environmental factors that, in earlier days, literally determined survival. Today, thanks to a wide range of state policies, such environmental conditions do not necessarily determine life and death, but they still have a significant impact on life chances and quality of life.

Second, the activities of work in such communities are constrained by the network of social relationships that constitute the basis of community life. Above all else, people in small communities must maintain reasonably amicable relationships with one another. Not only do they come into close contact every day, but they are invariably related to virtually everyone else, either by immediate family relationship or through intermarriage over several generations. Furthermore, in an often-harsh physical environment where few, if any, have a cushion of wealth to help overcome unexpected shocks and cope with vicissitudes, it is important to remain on friendly terms with those who will form a support system in times of need. It is almost inevitable that, sooner or later, everyone will need his or her neighbour's physical and social support. In terms of our focus on the fishery, the need to maintain a modicum of amity and cooperation constrains the extent to which a community member can engage in fishing activities that might create social rifts.

Third, life and work are constrained by the norms and values that form the basis of the community's distinctive local culture. To some, it may seem peculiar to talk about a community culture. Many social scientists appear to be preoccupied with 'proving' that there is a national culture and identity that supersedes any regional or local cultures. Consequently, regional and community cultural differences are often overlooked or denied. Yet, as the following community descriptions will make apparent, even the smallest communities may have a distinctive set of values, shaped by their environment, their history, and the external social, political, and economic forces that have influenced their development.

As the first of the two quotations that open this chapter suggests, the value system of some small Newfoundland communities is imbued with an 'ideology of cooperation and non-aggression.' Explanation in terms of ideology is always problematic. Ideology and values are often said to 'cause' social behaviour, but the source from which they themselves spring is often ignored. In the Newfoundland communities on which we focus here, the ideology of cooperation and non-aggression is, at least in part, a product of the need for amity that we have just described. None the less, that ideology is also an important element in the conditions of work in the inshore fishery. As we noted in the preceding chapters, it is widely assumed that a common-property resource will, by its very nature, bring about competition and conflict, and, as we argued in chapter 4, the main constraint on such conflict is a community culture that emphasizes cooperation and unity. Thus, the success of the inshore fishery depends in part on the extent to which an ideology of cooperation operates to constrain conflictual behaviour on the fishing ground.

Fourth, the activities of fishermen are constrained by a wide range of external organizations, agencies, and actors, including the state, the union, fish plants, fish buyers, and even fishermen in other communities. As we shall see shortly, much of the process of making a living in the inshore fishery involves dealing with the power and influence of such external forces. Indeed, these external agencies set the conditions within which strategies of work and strategies of earning a living are devised.[1]

CHARLESTON

Charleston (1986 population: 242) doesn't look like a Newfoundland fishing village. Most Newfoundland fishing villages cluster picturesquely around a harbour, with houses climbing to the churches and cemeteries on the hills above. The drive to Charleston from the Trans-

Canada Highway takes one through forty-five kilometres of hills and marshes that are locally termed 'barrens'. Finally, at the top of a long, high hill, one sees a glimmer of ocean in the distance. But this is no major harbour; it is a small inlet, one of the innermost arms of Bonavista Bay, located at a considerable distance from the open sea. At the bottom of the hill a marker indicates that Charleston is on the road that exits left. Yet a drive up that road reveals no community in the sense of a distinctive physical entity: Charleston has no community centre. It is a 'line' community, composed of a string of houses scattered along eight kilometres of road. Its most imposing feature is not a church or a community wharf, but a fish-processing plant located near the water about half way along the road.

Few Newfoundland communities can boast fish plants, and Charleston's is a relatively large one. To anyone familiar with other fish plants in other communities, this one instantly seems anomalous. Fish plants are usually located in prosperous fishing centres, typically situated near the mouth of a bay. This community and its fish plant are located virtually inland, far from the major fishing grounds. Few fish of any size are likely to reach this shallow inlet so far up the bay; consequently, few boats of any size are likely to fish here.

Despite these circumstances, government lists indicate that thirty-eight residents of Charleston are licensed to fish commercially; thirteen hold full-time licences and twenty-five hold part-time licences. In addition, several Charleston residents hold special licences enabling them to fish for protected species. There are twelve such licences (seven for herring, three for salmon, and one each for lobster and capelin), but they are held by only six people. Three fishermen have a herring licence; two have both a herring and a salmon licence; one has a herring and a capelin licence; and one has a lobster, a salmon, and a herring licence. The most sought-after licences are usually those for lobster and salmon, because the cash value of such fish is very high. Thus, the person who holds both a salmon and a lobster licence is lucky indeed.

The licensing lists for Charleston yield another, unusual piece of information. From the names of the licence holders, it would appear that at least six are women. All are part-timers. Although women have traditionally held important roles in the preparation of fish, few have become fishers in their own right. Indeed, it would be quite unusual to find six licensed female fishers in a large fishing centre; to find so many in a community as small as Charleston stretches the limits of probability. This fact alone leads one to anticipate a unique organization of fishery work in the community.

The following analysis is based on interviews with twelve of the community's thirty-eight licence holders. Among those we interviewed, seven held full-time licences and five held part-time licences. Two of the part-time licence holders were women. The sample was picked using the stratified random design described in chapter 1. The interviews lasted approximately three hours each, and provided an extensive insight into the fishing activities and general life of the community.

'I'm a jack of all trades. A little bit of sawing and a little bit
of fishing, and whatever else I get me hands on within the law.'
(Respondent no. 012)

Work in Charleston is not so much an occupation or even a pastime, as something to be hunted, tracked down, and pieced together. The community lives by the sea but owes its survival just as much to the nearby woods. The people of Charleston compensate for its locational disadvantage with regard to fishing by engaging in woods work. Yet, a closer look reveals that the 'fishermen' of Charleston are not descended from traditional 'fishermen-loggers'. Neither their logging activities nor their fishing activities fit traditional patterns.

As we noted in chapter 2, 'logging,' in the sense of being employed by a paper mill to cut timber, has become a full-time, unionized occupation. Those who engage in woods work in Charleston are not 'loggers' in that sense; they are better described as woodcutters. Some spend part of the winter cutting wood for sale as firewood; other have small sawmills (a glorified name for an old car engine attached by belts to a large saw blade), where they cut their own and their neighbours, logs into lumber for building houses and the occasional boat:

I has me own little sawmill here, and one in the woods. I saws wood for other people. (Respondent no. 012)

The way these fishermen engage in the fishery is also unusual. The staple of the fishery in Charleston is not cod or even lobster, salmon, or herring; it is squid. Most Newfoundland fishermen treat squid largely as bait, and catch it as a pastime on summer evenings after a day's work. However, the Charleston area seems to have had an abundant stock of squid in recent years. They have been so plentiful that many of Charleston's fishermen make a significant part of their annual income from squidding during the few weeks when these molluses are in the area. Indeed, as we shall see later, many continue to fish after the spring squidding season only to establish their credentials as full-time fishermen or to meet eligibility requirements for claiming unemployment-insurance during the winter months.

Their distinctive pattern of work activities is reflected in the way the fishermen of Charleston describe their lives. Only three of the twelve people interviewed, when asked to describe what they 'do for a living,' stated unequivocally that they were fishermen:

I've been fishing since the first day of April this year. That all I done since fours years ago. What we got has come out of the water. (Respondent no. 001)

Others tended to indicate a variety of occupations, referring to themselves as 'a jack of all trades' (Respondent no. 012), saying they were engaged in 'a bit of everything' (Respondent no. 002), or emphasizing the division of their labour between fishing and woods work:

I'd say that, while there's a fish to be got I'm at it, and when that's over its in the woods. (Respondent no. 005)

If you goes into the woods you're guaranteed to get a log, but you're never guaranteed to get a fish out of the water. That's why I be's mostly at the wood. There's always a tree there to get. (Respondent no. 012)

One person declared an involvement in only one type of fishing:

I works at the squids. (Respondent no. 011)

It was expected that those who held part-time fishing licences would be involved in some other form of work activity as well: some were 'doing carpenter work down at the fish plant'; others were engaged in trucking. Unexpectedly, however, some of the people who held full-time licences also identified themselves as other than full-time fishermen. Among them was one person who responded 'Right now I'm unemployed'; and another two, in their late fifties, indicated that they were retired and living on War Veterans Allowance.

The ambivalence of part-timers towards the fishery was reflected in their replies to questions about what they like most and least about it. Whereas full-time fishermen indicated that what they liked most was 'to be outdoors' and 'being me own boss,' part-timers exhibited a general dislike for a life on the sea:

I wouldn't like to be on the water too much. I'm no lover of the water. (Respondent no. 008)

I don't like the open high seas. The only thing I been out in is the open boat. (Respondent no. 012)

Perhaps the most unexpected finding was one relating to the fishing activities of the two women in the sample. Both were 'fishers who never fished.' While their husbands were engaged in catching squid, these women were employed on shore, drying the catch:

My husband catches the squid and I drys them. I'm rarely if ever in a fishing boat. There's not too many women who gets [unemployment insurance] stamps for squids be's out in the boat. You got to have somebody ashore. If you're out there and a shower of rain comes on, that's it! You might as well stay out there. (Respondent no. 002)

I'm a housewife. I works at the squids. I'm waiting for the squids to come now. (Respondent no. 011)

The disproportionate number of women fishers in Charleston thus proved not to be an indicator of increased female involvement in catching fish. Rather, these women were employed in a combination of activities – drying fish and housework – that was once the accepted work pattern of rural Newfoundland women.

In one particular way, however, these women's role in the fishery is new and should not go unnoticed. Indeed, if this new role for women becomes more widespread, it has the potential to transform the inshore fishery. The significant aspect of these women's activity is revealed in one of the preceding quotations: They are not fishing for fish, but are engaged in accumulating unemployment-insurance stamps. In the past, women's economic contribution to the fishery went officially unrecognized. Most women spent many hours of each summer day spreading salted fish to dry in the sun (and retrieving it at the first hint of rain).

Although they added a considerable economic value to their husbands' catch, they received no money for their labours. Their work in the fishery, like their housework, was not recognized as an economic activity. In part as a result of this, women in most communities gave up drying fish when the fish-processing plants were introduced. Their husbands now sold the fish fresh, and the women worked for wage labour in the plants. With only a few weeks of such labour, they became eligible to collect unemployment-insurance throughout the remainder of the year. Dried squid, however, has a high cash value, which has prompted some women in Charleston to return to drying. Some families in the community are still satisfied with having the husband sell the squid, and collect unemployment-insurance stamps in his own name. Others have discovered that it is possible for the wife to obtain a part-time fishery licence to dry and sell the squid in her own name, which also entitles her to collect the precious unemployment-insurance stamps that guarantee an income during the winter. If the husband in such a household can catch and sell enough fish, or find additional land-based work, to qualify for unemployment insurance, both spouses can collect benefits throughout the winter – without the wife's ever having to leave home.

This practice is considered by some Charleston residents to be unethical, immoral, and illegal. However, salt fish (of any species) is now something of a rarity, and consequently brings quite a high price. This suggests the possibility of a new role for women in the fishery, on the model of the involvement of the Charleston women. The latter is one example of how the residents of Charleston have adapted state policy to their own ends in ways that were undoubtedly unanticipated by the policy makers. It is also an example of how the process of earning a living has become a 'calculated activity' in this community, as well as in many others. We shall return to both these issues later in the chapter.

'First there, first served. A lot of people has their traps in the same place each year.'
(Respondent no. 001)

The potential for conflict is inherent in the fishery; it stems from the nature of its property relations and access rights. Consequently, the way fishing grounds are allocated can have a critical effect on community solidarity. Because the primary activity of Charleston fishermen is squidjigging, one would assume that there was less potential for conflict here than in other communities. Property rights are not much of a factor in squidjigging. Once squid 'strike in,' they are relatively abundant and squidjigging boats cluster together over the largest schools. This is perhaps why, at least according to eleven of the twelve Charleston residents we interviewed, the community has no formal system for allocating fishing berths. One respondent, when asked if his community had a way of controlling where nets and traps were set in the water, replied:

No, but there's not a lot be's at it here. They [fishermen] don't be out that much. In spring. But now when the squid comes, they gives it [fishing for other species] up. (Respondent no. 012)

Another's response gave evidence of the 'ideology of cooperation' that is central to the mythology of many fishing communities:

We does everything peaceful in this area. We don't set [out nets] across another fellow's nets, and we don't take another person's place. I has my nets were I fished as a boy. [I use] marks that was given to me by my forefather – the marks for Keough's Rock, Offer Rock, Matthew Way's Point. But, if someone was on my place I'd go somewhere else. (Respondent no. 003)

One fisherman, while expressing the view that there was no customary regulation, also indicated that most people tended to use the same fishing locations from year to year.

First there, first served. A lot of people has their traps in the same place each year. (Respondent no. 001)

The sort of system our respondents described works as long as all fishermen recognize the rights of others to remain in their traditional fishing spots. And since several fishing communities may occupy the same fishing ground, a similar attitude is required of the fishermen from nearby communities. Finally, such a system is likely to work only when overcrowding is not a problem – in other words, only when there are fewer traps than desirable locations.

Other respondents provided information suggesting that the traditional approaches are already beginning to break down as a result of increases in the amount of equipment being used and in competition from other communities. Several respondents noted the recent increase in the number of fishing traps being set:

This is the first year we've [i.e., his crew] had traps. I'd say there's a dozen or more new traps around here this year. There's at least four more from this place alone. (Respondent no. 001)

One respondent described two recent conflicts over the allocation of fishing sites. Both were disputes between Charleston residents and residents of a nearby community:

A fellow here put out his mooring first to mark the spot till the fish come. Someone else from [a nearby community] put his trap there, right under the other fellow's moorings. Two sets of moorings there and only one trap.

We had a cod net in a trap berth and a fellow from up the shore phoned Saturday evening and said he wanted to put a trap there on Monday. We went out Sunday and he had taken 'em [respondent's nets] out of it; just tossed 'em off, and put his there. The same guy moved a salmon net belong to another fellow and he got to go to court about that. In [a nearby community] they does it all the time down here. They's a hard crowd. Always into it over trap berths. (Respondent no. 010)

In addition to providing evidence about conflict, the latter example establishes that there are traditionally accepted locations seen specifically as trap berths, which are not to be occupied by other types of gear. So strong is this custom that the fisherman in the example felt he was within his rights to ask to have cod nets removed from one such location so that he could set his trap there. This is an indication of conflict over gear as well as community regulation of gear, which will become a dominant concern in our analysis of fishing activity in larger communities.

'There's nobody knows what a part-time fisherman is.'
(Respondent no. 010)

The pattern of work in Charleston would be much as we have already described it, were it not for the influence of state policy. State policy impinges on that work world in a variety of ways, shaping and moulding its structure and altering the activities of the workers themselves.

The most basic way in which it does so is through the licensing process.

The primary distinction in licensing is that between full-time and part-time fishermen. According to official definition, full-time fishermen work year round without involvement in other employment, while part-time fishermen engage in other employment activities in addition to the fishery. Some of those we spoke to clearly accepted this distinction.

Yes, the full-time fishermen are out there all of the time. (Respondent no. 011)

Full-time fishermen are at it for a living. I'm a logger – full-time in cutting logs and collecting unemployment insurance. (Respondent no. 006)

Ironically, all of those in Charleston who accepted this definition held part-time fishing licences. Those who held full-time licences presented a far more complex picture.

The difficulty with the full-time/part-time distinction lies in the fact that the fishery is a seasonal occupation in most areas of Atlantic Canada. Few full-time fishermen actually do fish full-time, and couldn't do so even if they wanted to. Weather and ice conditions, as well as the seasonal availability of fish stocks, make full-time fishing impossible. The reality of the situation, then, is that most full-time fishermen fish part-time. The irony of this was not lost on most of our respondents.

Lot's of fellows [who have] got part-time licences, holds other jobs. Some fellows holdin' full-time licences got other jobs too ... There's a lot of people with full-time licences shouldn't have 'em. All they's at is squid. (Respondent no. 001)

There's a lot of full-time fishermen, when it comes cold, who are not too fussy if they goes out or not. [On the other hand] there's a lot of part-time fishermen like myself fished till after Christmas last year. (Respondent no. 003).

Take a fellow full-time fishing in Bonavista – every cent he makes, that's where he make's it [i.e., from the fishery]. Around here, soon as fishing gets bad they be's at something else. That's not a full-time fisherman. (Respondent no. 007)

There are more rows over a part-time and full-time licence than anything. If a fellow got 'em [a full-time licence], I don't tell him he shouldn't have 'em. But in my mind he shouldn't have 'em. (Respondent no. 001)

The situation is complicated further by the system of allocation of protected-species licences. Such licences are often in the possession of fishermen who hold only a part-time licence. Some are elderly fisher-men who, while holding on to their lucrative species licences, have given up the heavier and less financially rewarding tasks associated with full-time fishing. The consequence, however, is that many suppos-edly full-time fishermen cannot get the species licences that would allow them to make an adequate living from the sea. As a result, they find it necessary to engage in additional labour outside the fishing industry.

In Charleston this system of allocating protected-species licences has produced a level of hostility and animosity that makes something of a mockery of the myth of community cooperation and amiability:

There's nobody knows what a part-time fisherman is. There's part-time fisher-men with lobster, salmon, and groundfish licences, and they don't go out in a boat. Oh, they might set a salmon net or two to get one to eat, but that's all they do. There's only four of us here fishing full-time. Only one got a salmon licence. None of us got a lobster licence ... A full-time fisherman should be able to get into it all [i.e., all species] and make a living at it. I'm classed as a full-time fisherman and all I can do is catch a few codfish and squid ... There was some here last year didn't even own a boat who had a full-time licence. Then, there is some here with just a part-time licence have a salmon and lobster licence. (Respondent no. 010)

Well that is what I am [i.e., a full-time fisherman]. But I am a full-time fisherman and I can't get a salmon licence. On the other hand, people with a salmon licence can't get a [full-time] fishing licence ... You should be able to fish anything to make a living out of it. (Respondent no. 009)

I don't see the difference between a full-time and a part-time licence under the law as it is today. Full-time fishermen should be given all licences for every protected species. The one licence [should] cover the works. The part-time fishermen should be given the licences for groundfish and squid. That would be fair. I haven't got a lobster or salmon licence so I'm not a full-time fisher-man. (Respondent no. 003)

There's a fellow here fishing 500 lobster pots single-handed, making $25,000 or $30,000. It'd be better to divide the licence up between two or three fellows with 100 pots apiece. Then they could all get some lobster. Now one fellow gets a big haul and the others have to sit ashore and do nothing. (Respondent no. 005)

Some of the hostility was directed at holders of protected-species licences who used them in the spring lobster and salmon season, but then left to take up seasonable wage employment outside the community.

There's fellows from down the shore have both a salmon and lobster licence. Then they goes on the lake boats. Then they comes back and draws their unemployment [insurance]. (Respondent no. 001)

Even greater hostility was directed at holders of lobster or salmon licences who did not use them as fully as was permitted, spending much of the lobster season outside the community, engaged in other work.

[He's] got 500 lobster pots but only spent a week at it this year, and last year he was up north working. He got nineteen lobster before he went up on the boats. (Respondent no. 010)

There's a lot of old fellows now with a lobster licence for 150 pots, and they only use 20. A fellow not using something shouldn't have it. (Respondent no. 007)

The frustration caused by this loss of income to the community was exacerbated by the obstacles the system placed in the way of transferring species licences from one person to another. Most respondents clearly believed that licensing regulations should allow them to pass licences from father to son, and perhaps also between brothers or close friends. They noted that sons often took over ownership of their father's boats and gear, but could not make an adequate living from them unless they could also acquire the requisite licences. They argued that, without such rights of transfer, there was no incentive for fishermen to give up licences that they were not fully utilizing. The strength of the respondents' convictions on this point is evident in the vehemence of the following statements:

Jesus, the thing is a hard racket to figure out. I passed me sawmill licence over to me brother because we heard that anyone with a mill licence wouldn't be able to get a fishing licence. It was only rumour. If the old man gives it up [i.e., a protected-species licence], he should be able to give it to his son. (Respondent no. 012)

There's a fellow down here tried to give his licence over [to someone] and they wouldn't let him. If he did, at least someone else would use it. It would always stay in the place and be used. (Respondent no. 001)

There's a fellow in [a nearby community] wanted to give his [salmon licence] to me and he couldn't give it. That's a licence for this year has not been used. Also [an elderly Charleston resident] would give his licence to me, and he can't. He just wants a few [salmon] for himself. But it's foolishness for him to give it up and everyone lose it. But he would give it to us if he could. (Respondent no. 005)

I can't see why you can't pass it on to someone else. If you got a sawmill licence, you can pass it on. (Respondent no. 008)

Such comments clearly indicate that Charleston fishermen generally regard licences as a form of property – specifically, as family property, to be passed on from father to son, or to other relatives. However, there is also a suggestion that licences are seen as community property, a community resource that should remain in use within the community. The respondents' references to sawmill licences are instructive in this regard. Although the sawmill, like the boat, is private property, the trees are on Crown land and are state-regulated common property. Yet licences to harvest trees are readily available, while licences to harvest certain species of fish are not.

'There's lots of full-time fishermen fish till they gets their stamps,
and then don't give a damn if they goes on the water.'
(Respondent no. 003)

The licensing system represents only the most obvious and immediate way in which state policy impinges on the work activities of inshore fishermen. It does so by other means as well, the most significant of which is the unemployment-insurance system. Whereas the licensing system establishes who has the right to fish, the unemployment-insurance system dictates how much annual income a person can receive from his or her fishing efforts. However, the two state policies are not entirely separate. Possession of a licence often determines whether or not one can sell one's fish; eligibility for unemployment insurance is determined largely by the amount of fish one has sold. Before examining how this system actually operates in Charleston, a brief description of the unemployment-insurance system is in order.

The Canadian unemployment-insurance system is a government-run program designed to assist Canadians who are temporarily unemployed. Employees and employers make regular contributions, proportional to the employee's wages, to a central fund. Should the worker become unemployed, he or she may apply for compensation from this 'insurance' fund. The amount received in benefits will depend on the number weeks of previous employment, the wages earned, and the region in which the worker lives. Recognizing that employment opportunities are scarce in some parts of Canada, the federal government has regional guidelines establishing a ratio between the number of weeks worked and the number of weeks for which a claimant is eligible for unemployment insurance. Newfoundland, being one of the poorest regions of Canada, has the most favourable ratio. At the time these data were collected, ten weeks of prior work entitled the unemployed worker to benefit over the next forty-two weeks.

Like other workers, fishermen are entitled to unemployment insurance, but they are subject to somewhat different regulations. When the government extended the unemployment insurance program to fishermen, it was decided that benefits would have to be tied to income from fishing, since there was no other way to verify that a claimant had indeed worked during the fishing season. A system using different values of unemployment insurance stamps was instituted. Every week that a fisherman sold fish to a licensed buyer, he received a stamp, which was pasted into his stamp book. The stamps had different values, proportional to the value of the fish sold during that week; in other words, the greater value of the fish sold, the greater the value of the stamp received. To be eligible for unemployment-insurance during the following winter, a fisherman had to have sold fish for at least thirteen consecutive weeks in the calendar year. The amount of benefits to be received varied with the total value of accumulated stamps. Those with high value stamps received considerably more unemployment insurance than those with low value stamps. Unlike other workers, fishermen are entitled to receive benefits only between January 1 and the end of May, regardless of the length of previous employment. Furthermore, their benefits can be discontinued if they begin fishing and receive stamps at any time during that period.

This system involves the fisherman and his family in a series of 'gambles'. The dilemma he faces begins in the spring, with the decision about when to start fishing. Should he begin before his unemployment insurance runs out? For those who hold salmon and lobster licences, this is a major dilemma. Lobster and salmon are often most abundant during the late spring. These species have such a high cash value that, in addition to bringing in a good income, they can also ensure a very high value unemployment-insurance stamp, to count towards the next winter. But spring is also the time that floating ice packs are at their worst. The fisherman may not be able to put out his nets and pots, or he may have them destroyed by ice. Moreover, if the fishing is bad, he may not get as much money in fishing income as he would have received in unemployment-insurance benefits. If that occurs, he will be further penalized during the following winter, as his first two or three weekly stamps will also be of low value, and they are a significant portion of the amount on which his unemployment insurance for the following winter will be calculated. Unemployment-insurance regulations allow only the first thirteen stamps received by a fisherman to be counted. In other words, even an astonishingly good catch in the latter half of the season will not help increase his unemployment insurance for the following winter.

The temptation to begin fishing early may not be as great for fishermen who do not have lobster and salmon licences. None the less, by mid-May, they still have to decide whether to give up unemployment insurance in order to start catching the cod that are available. The stamps they receive for them may be of low value, but they are better than no stamps at all: Failure to collect stamps for thirteen weeks would result in *no* unemployment insurance during the following winter. Many a fisherman has waited until the end of May to start fishing, only to discover that the schools of fish do not stay in the vicinity of his community for an additional thirteen weeks.

Faced with such economic uncertainty, fishermen endeavour to ensure as stable and high an income as possible. To understand the process this involves, it is useful to consider the differences between fishermen and farmers. Farmers are harvesters who plant a crop and tend it while it grows. Fishermen, by contrast, are hunters and gatherers. They hunt their prey, and their survival and success depend largely on their knowledge of the environment and on their imagination and skill. The point is that the 'psychology' of fishing is considerably different from that of farming, and has implications for the way fishermen confront other aspects of life as well. This is all the more true for part-time fishermen, like many of those in Charleston. They approach their world as hunters, looking for ways to track down and capture a wide variety of sources of income. Unemployment insurance and other forms of state assistance, no less than the fish and the forests, are among those sources of income.

'It's like a game of cards now. Except everybody makes up their own rules.' (Respondent no. 012)

The preceding quotation compares the Newfoundland inshore fishery to a game of cards. As we noted in chapter 3, the rules that emanate from the state set the parameters within which that game can be played. However, as we also noted earlier, in addition to hunting fish, fishermen must also hunt the income to be derived from the sale of those fish. It is important to recognize that the two processes are not synonymous. While it is state regulations pertaining to licensing that determine the way fishermen hunt fish, it is state policy on unemployment insurance that has the major influence on the strategies they adopt in their search for income.

We have already discussed one income-generating strategy used in Charleston, namely, having fishermen's wives acquire licences to sell dried squid. This strategy allows a wife to sell squid over a thirteen-week period in order to get her own unemployment insurance. Her husband will also sell squid, either 'green' or dried, which will count towards his own thirteen weekly stamps. As one fisherman explained:

We gets squid stamps in whoever's name needs it. Last year I shipped some in my own name too, just so it wouldn't look bad. What I sells though is mostly green. (Respondent no. 002)

Still, not everyone is happy with this system. One fisherman complained bitterly that, in order to get stamps, he and his wife were forced to sell squid to the fish-processing plant during the fishing season, when the going rate was low. (A higher rate might be available later in the year from private buyers.) He felt that it was unfair that part-timers, whose unemployment-insurance benefits were secured by their ten weeks' work outside the fishery, were able to hold on to their squid and get the more favourable price.

Part-timers can hold on to squid and get a good price after the price went up. We had to sell ours in order to get our stamps. Them with a part-time licence, all they're waiting for is the squid. I hope to Jesus they never strikes in. (Respondent no. 001)

Probably because the drying of squid has brought women back into the fishery labour force, some families in Charleston are also beginning to salt and dry cod in the traditional manner. High-quality salt cod is now in considerable demand and commands a high price. This strategy can therefore add labour value to the fish sold and bring increased income into the family. Because the wives who dry the cod usually have enough income from their dried squid to qualify for unemployment insurance, the cod tends to be sold in the husband's name. This raises both the husband's income and, potentially, the value of the unemployment-insurance stamps he obtains.

One fisherman, however, noted an undesirable consequence of combining fresh-fish and salt-fish stamps. He attempted to combine ten weeks of fresh-fish stamps, two weeks of salt-fish stamps, and five weeks of stamps from working for a road-construction firm. Instead of totalling this to seventeen weeks, the Unemployment Insurance Commission (UIC) ruled that each salt-fish stamp represented more than single week's work, and credited him with twenty weeks' work. Ironically, the effect of this generosity was to lower his level of income for the crucial first thirteen weeks on which the fisherman's unemployment-insurance rate was based. In the fisherman's words:

They called it twenty weeks just because they wouldn't have to pay me so much. I didn't get paid [i.e., unemployment insurance] for my salt fish. But it lowered the average for my other fish. (Respondent no. 005)

As a result of this practice, other fishermen in Charleston who prepare salt fish often sell it to buyers who refrain from issuing unemployment-insurance stamps.

I sold my salted cod to a co-op in Clarenville that doesn't give stamps. (Respondent no. 009)

In Charleston, few fishermen have the option of basing their income-generating strategy entirely on their fishing effort. The local fishing grounds are simply not productive enough to make this a viable option. As a result, most Charleston fishermen find it necessary to acquire what they call 'land stamps,' earned for land-based work, in addition to their fishery stamps. There are decided advantages to this strategy, because land stamps, unlike fishery stamps, are good for unemployment insurance year round. There is even a formula that allows a small number of fishery stamps to be used in conjunction with land stamps to qualify for regular unemployment-insurance benefits. As a result, many Charleston fishermen do whatever they can to maximize the number of land stamps they can acquire. As one respondent put it, 'You got to make a dollar where it's to be made' (no. 005). Some derive part of their income from their woods work and their sawmills, but this is rarely sufficient to qualify them as land-based workers.

You can't get stamps logging, and there is no way in the world you can get twenty fishing stamps around here. (Respondent no. 012)

Others quit the fishery as soon as the lobster, salmon, or squid season is over and take up work outside the community. But full-time fishermen who adopt this strategy run the risk of being reclassified as part-time fishermen.

You can only lose a month from the fishery and you got to fish two years like that to become a full-time fisherman. (Respondent no. 005)

One person's licence was reclassified from full-time to part-time status after he spent three months working away from home. He appealed the decision and had it reversed, based on evidence that he had begun fishing through the ice in mid-February, and had therefore, fished the necessary thirteen weeks before taking up his outside employment. Another person with a trade that enabled him to work outside the community indicated that he was frequently approached by outside companies offering him work. Although he could have made a sizeable income with such a company, he chose instead to work only a month or so a year in his trade so that he would not jeopardize his status as a full-time fisherman. He preferred fishing because it allowed him to return to his home and family every night:

I work at home and I'm home every night. On construction you might be in Corner Brook or Manitoba and away four or five months ... I see fishing getting a bit better. You don't want much fish now to make a day's pay. (Respondent no. 005)

Many of the fishermen in Charleston benefit from the presence in the community of an entrepreneur who, in addition to his activities as a part-time fisherman, operates a small business that includes five or six trucks. He hires some community residents as part-time drivers, and some to cut wood, which he trucks to the city for resale. He also buys squid and fish from local residents and trucks them to plants in regions experiencing shortages. This operation represents a sort of additional insurance policy for many Charleston residents. When they are short two or three stamps to qualify for unemployment-insurance benefits, they can make them up by performing a couple of weeks' work for this firm. We frequently heard comments such as the following:

I sold my dried squid to [the local entrepreneur] because he was paying a better price. (Respondent no. 010)

I got three stamps in the winter cutting sticks for [the local entrepreneur]. But this wasn't for this past winter. I did the work the winter before. He paid me $800 and I got $292 unemployment insurance every two weeks clear. (Respondent no. 007)

Finally, a significant part of many Charleston residents' income is from the direct sale of goods and produce. Firewood and lumber are sold for cash. At least two of our respondents built and sold boats during the preceding winter. Fish in excess of what is needed to get stamps is sold for cash to fish buyers and supermarkets in nearby centres. We will examine some of the implications of such sales in the next section.

'The fishery – it all depends on the plant.' (Respondent no. 001)

Although the preceding discussion gave close attention to the role of the state in affecting the work and income-generating strategies of Charleston fishermen, it largely ignored the other important external agencies that influence the nature of work in Charleston. These include the fish plant and other fish buyers, and the fishermen's union. There are areas in which there is a conflict of interest among these agencies, and between them and the state. It is the argument here, however, that, in many ways, these other agencies also combine their efforts to provide a framework of regulations and social relationships over which the fishermen of Charleston have little control, and within which they must operate.

Of these other agencies, the fish-processing plant probably plays the most important role in the work and income-securing activities of Charleston. Not only do many of the residents work there, but many of those who fish sell their catch to the plant. Some respondents emphasized the importance of the plant in their lives:

According to local talk, I hear about it closing down. Probably nothing to it. I hope it doesn't. If the plant here closes, I don't know what people here would do. (Respondent no. 011)

The only work here is the fish plant. When the plant closes the business is gone ... The fishery – it all depends on the plant. The plant had a like to close this year, but the government stepped in and gave 'em some money. (Respondent no. 009)

The fishermen's union also plays a significant role. The union contract with the plant requires that, in buying fish, the plant give preference to full-time fishermen over part-time fishermen. Consequently, during the peak of the season, part-time fishermen may find themselves unable to sell their fish. Indeed, some perceived this to be the only real difference between full-time and part-time fishermen:

The only difference as I see it is that a full-time fella can sell his [fish] sometimes, and a part-time fella can't sometimes. (Respondent no. 008)

However, there are several other fish-processing plants within eighty kilometres of Charleston, including those in Summerville, Bonavista, Arnold's Cove, and Port Union. They give Charleston fishermen a variety of options when it comes to selling fish. Furthermore, because the Charleston plant is situated on a shallow inlet, many fishermen cannot unload the fish they sell to the plant from their boats. Instead, they must load it on a truck and drive it to the plant. In addition, the Charleston plant reportedly lacked some of the equipment available in other fish plants to facilitate the unloading of fish:

Summerville's got good equipment for handling fish. They have an ice machine, conveyor, and even nets for taking fish out of boats without having to prong it. Therefore they can give top price. To sell fish to Charleston you have to have a truck. It is more work and the quality is poorer because of the way they handle it. (Respondent no. 010)

Three years ago a Russian vessel bought squid and it was a real comfort compared to the plant. Because in the plant you got to get on the wharf and truck 'em up to the plant. To sell squid to the plant here you got to have a truck to truck it to the plant. (Respondent no. 009)

Once the fish are on a truck, there is little to stop the fishermen from driving the short distances to other plants in the area. In addition, there are several private buyers in the region who speculate in fish, buying it up and trucking it to the fish plant that is offering the best price on any given day.

He [a fish buyer] buys fish and trucks it to any plant that needs it ... Bonavista, Charleston, Arnold's Cove. He knows at the beginning of each day which plants will take how many truckloads from him. He buys this up and moves it around. (Respondent no. 001)

Selling some or all of their fish to these alternative outlets is one way in which Charleston fishermen try to limit the power that their local plant has over them.

Faced with such competition, however, the Charleston plant has countered with regulations of its own. For example, it will buy squid only from fishermen who sold their cod to the plant the previous year. In addition, it sometimes places a quota on the amount of fish it will buy from any one fisherman. These regulations were highly controversial in Charleston, and had a significant impact on its residents' strategies for making a living:

I come in and I could only sell 300 of the 1300 pounds [of squid] I had. I couldn't make nothing at that, so I was drying my squid last year. (Respondent no. 010)

Last year I didn't ship [i.e., sell] any fish [i.e., cod] there. They bought a couple of lots of squid. Then one morning I had 1000 pounds and they wouldn't buy it. That's ridiculous ... I had a fish licence and a boat licence, so I should be a fisherman, eh? ... That's a rotten Jesus thing they done last year. (Respondent no. 012)

Last year [the plant] put us on a quota of 500 pounds a man. I went to the manager of the plant and told him I couldn't survive on $30 of squid a week and there wasn't no codfish. I told him to give them with a pension [a limit of] 250 pounds and give me more 'cause I needed it to live. No, he couldn't do that. So I told him I had to go away to work. I had to take me nets out of the water and went to work with [a construction firm] and came back Labour Day. I wasn't fussy about doing construction work as I would lose my fishing licence. That wasn't right. (Respondent no. 005)

One might expect that the fishermen so affected would take their case to the union. But none of the people interviewed gave any indication that this was an option they would consider. Indeed, they were highly ambivalent about the union.

It might help some fellows but it don't help everybody. (Respondent no. 012)

I suppose it helped some, but they're still complaining. (Respondent no. 11)

They haven't helped fellas like me [i.e., part-timers with other jobs]. (Respondent no. 006)

Most did praise the union for obtaining higher prices for fish:

Wouldn't get the price for fish without the union. I don't know if they've gone too far with it though. (Respondent no. 004)

They was selling squid here for six cents a pound. Now the union gets them a good price. (Respondent no. 007)

There was some resentment, however, about being forced to belong to the union, about having to pay dues for membership in it, and about some of its negotiating tactics with the plant:

The union is just a money-making racket. Millions of dollars going in and I don't see much for it. (Respondent no. 003)

Some respondents were highly critical of the union's impact on work in the plant:

The [plant] workers are protected too much by the union. I can't see it. The company is not getting its money's worth and it can't survive. (Respondent no. 003)

With prices of fish it's been a help. But a union can go too far. There's only so much a company can pay and that's it. (Respondent no. 005)

Overall, however, there were no strong feelings for or against the union. One respondent explained that he belonged 'because it's there.' Another declared, 'It's just as well to be in it. They're taking out union dues anyhow' (Respondent no. 002).

'There's no better place.' (Respondent no. 004)

Despite the various strategies and calculations required to earn a living, there can be no doubt that the residents of Charleston like where they live. When asked whether they liked living there, all twelve respondents answered emphatically in the positive (see Table 5.1). Like Newfoundlanders in many other communities (see Matthews 1976), they were convinced that there was 'no better place' for them to live:

Jesus, I loves it. I don't think I feel content anywhere else. I knows every patch of woods back of here. (Respondent no. 012)

Similarly, all twelve responded that, as far as they were concerned, Charleston was a good place to raise children. They were also unanimous in their positive assessments of fellow residents, whom they described as hard-working. The majority agreed that the people of the community 'cared about the place,' and although there was little support for the statement that the leaders of the community were good and capable, this was not a negative evaluation as much as an indication that the community lacked any formal or informal leadership structure. In the words of one respondent,

Everyone is handy about equal around here. Most of life they've been their own boss, and they don't like anyone looking after them. (Respondent no. 012)

They were almost as positive in their evaluation of the services that the community provided. Most felt the schools and the teachers were good and the medical services adequate. Few thought they had to do without a lot:

We're not without a thing in this place. (Respondent no. 007)

TABLE 5.1
Charleston respondents' views of their community and its future

	Yes	No	Uncertain
Evaluation of community			
Like living here	12	0	0
Good place to raise children	12	0	0
People here generally hard-working	12	0	0
Leaders are good, capable people	2	1	0
People here care about the place	10	1	1
Evaluation of community services			
Good schools and teachers	10	2	0
Good medical care	8	1	3
Enough to do in spare time	4	5	3
People here do without a lot	3	8	1
Evaluation of employment opportunities			
Good place to find work	1	10	0
Fishery has a future here	6	1	5
Fish plant here has a future	3	2	7
Inshore fishery good for young people	6	5	1
Evaluation of community's future			
Children should settle here	8	1	3
Community has a good future	10	1	1

Note: There were 12 respondents. Where responses for any given item do not total 12, the balance represents abstentions.

Most respondents thought their community had a good future, and the majority felt that their children should settle there.

Real dissatisfaction became apparent only in the respondents' evaluation of employment potential in their community. Most agreed emphatically that Charleston was not a good place in which to find work. However, they were divided on whether the fishery had a future and whether it was a good occupation for young people. As we have noted elsewhere (Matthews 1976), rural Newfoundlanders often do not judge the quality of life in their communities exclusively in terms of employment opportunities. Indeed, the people of Charleston are very attached to their community; their problem is simply that there is not enough work for them to do there.

KING'S COVE

Unlike Charleston, King's Cove (1986 population: 255) fits the postcard image of a traditional Newfoundland outport perfectly. The community clusters around the end of a long harbour, and extends along its western side. At the entrance of the harbour stands a majestic old lighthouse. Halfway up the steep western bank, the spire of a massive old wooden church rises as a different sort of beacon. Many of the older houses are of a style unique to the Bonavista Peninsula and lend a character to the community not found in many other places. These large, two-storey dwellings with high pitched roofs attest to the former prosperity of the community and its fishery. The merchants of most Newfoundland communities could not have afforded such dwellings, yet in King's Cove, these were the houses of humble fishermen. The harbour itself is open to the sea and provides little protection from the ravages of a northeast wind. Today, however, the end of the harbour is protected from the ocean by a massive stone breakwater. In the boat basin inside, fishing boats crowd and jostle along the protected side of a new government wharf.

Over the hill on the other side of the harbour is a large new regional school. From here, the view is not of the sea, but only of the road leading out of town. The school is the educational centre of the region, and children are bussed in from many nearby communities to attend. Clustered along the highway near the school are several new, suburban-style bungalows, most of them the homes of the schoolteachers. Their modern style and their location away from the sea also symbolize a new orientation developing within the community.

King's Cove has a long and distinguished history in Newfoundland. According to Prowse's classic history (1895, 537), it was at King's Cove that John Cabot raised the Royal Ensign to claim possession of a new continent for the King of England, and thereby, gave the community its name. It is uncertain exactly when King's Cove was first settled. However, Bonavista, some twenty kilometres by sea to the north, and Keels, the next harbour to the south, are recorded in the 'List of Planters' Names' for 1676, and it is probable that King's Cove was settled around the same time (Macpherson 1977, 107–8). Certainly, the community must have been well established by the late eighteenth century, since many of the residents of nearby communities at that time are known to have migrated from King's Cove (Macpherson 1977, 114–28). Thus, there can be little doubt that King's Cove is one of the oldest continuously occupied communities in North America.

King's Cove is unusual in that it is primarily a Roman Catholic community – one of the few Roman Catholic fishing centres on the northeast coast outside the St John's–Avalon Peninsula area. In a province where relations between the descendants of Irish Catholics and of Protestants from the south of England have often been stormy, and where denominationalism still has influence, this feature has served to set the community apart.

Fisheries licensing records show that there are forty licensed fishers in King's Cove, thirty-two men and eight women. Of these, only ten hold full-time fishing licences; the remaining thirty hold part-time licences. All eight women hold part-time licences. Although the community is lucky enough to have eighteen protected-species licences (eight for salmon, eight for lobster and two for herring), they are distributed among only eight people (six hold salmon and lobster licences; two hold salmon, lobster, and herring licences). Significantly, two of the holders of the coveted salmon and lobster licences are part-time fishermen.

It is interesting to compare these data with those for Charleston. There are actually fewer full-time fishermen in King's Cove than in Charleston, and the number of part-timers is only slightly higher. This is startling, as Charleston has virtually no fishing history, while King's Cove has long been an established fishing community. In some respects, the licensing data for the two communities are similar. Both have a high percentage of part-timers, as well as a significant proportion of women holding fishing licences. As the communities are only about fifty kilometres apart, some similarity is to be expected. But King's Cove is well out towards the mouth of Bonavista Bay, near prime fishing areas, as well as an area noted for abundant lobster beds. One would therefore expect the nature of this community's fishery to be somewhat different from that of Charleston's.

We interviewed ten of King's Cove's fishers, including six full-time licence holders and four part-time licence holders. Two of the part-time licence holders were women. Six of the ten people interviewed held salmon and lobster licences and, contrary to the licensing data, all claimed to hold herring licences as well.[2] One of the protected-species licence holders held a part-time general fishing licence.

One cannot remain long in Kings Cove without becoming aware of the residents' high regard for their history. Quite often, respondents 'located' themselves with reference to the history of their community:

My grandfather and my great grandfather were the first settlers of King's Cove in 1750. (Respondent no. 021)

In the 232 years that the settlement has been here, there was only four years that an Aylward wasn't fishing. (Respondent no. 023)

Others were quick to list the distinguished Newfoundlanders who were born in King's Cove and had achieved success outside the community. The list includes the founders of several well-known business firms in Newfoundland, as well as former politicians.

The focus on the past and the interweaving of personal and community history were very much a part of the responses of the older fishermen we interviewed. Their biographies give a vivid picture of life in King's Cove over the past seventy-five years:

I started fishing in 1929, and my brother and I were fishing in a paddle boat [i.e., a row-boat]. We brought in our fish and split it and dried it ... We had our own wharf and premises, and now there's not a thing left down there. Every number of years it was swept away by a storm. I saw it [happen] in 1921 and my father built it up single-handed. We had to go into the woods and cut lumber and start it over again in 1948, and again in 1959. It was almost a periodical thing. (Respondent no. 022)

Our interest here is, indeed, in the way things have changed from the past to the present. Just as the new breakwater and government wharf have put an end to the 'periodical' destruction of boats and fishing premises, so, too, have new state regulations affected fishing practices in King's Cove.

'It be the only thing I does.' (Respondent no. 027)

Unlike the fishermen of Charleston, the fishermen we interviewed in King's Cove tended to identify fully with the fishery. They did not describe themselves as 'jacks of all trades,' but declared instead that the fishing was 'about all I be's at' (Respondent no. 020). Others made similar comments:

That's me occupation. I'm fishing full time. (Respondent no. 013)

I made no money, not a nickel for a number of years now, only [from] fishing. (Respondent no. 018)

The only exceptions were the two women respondents, who described themselves as 'housewives' who were 'just drying some squid, that's all' (Respondent no. 015), and one person who described himself as 'unemployed and getting unemployment insurance now' but also 'jigging squids six or seven hours per week' (Respondent no. 019). Surprisingly, that person held a full-time fishing licence.

A similar picture emerged when we asked the fishermen what they liked most about fishing. Their replies reflected their love of the sea and the value they placed on the freedom of choice that characterized their work:

It is a free life but you have to be experienced. You are your own master and sometimes that is the hardest boss you can work under. (Respondent no. 022)

It be my way of life. I likes everything. I likes to be on the water. It's real healthy. (Respondent no. 018)

I just likes to be at it. (Respondent no. 020)

Indeed, one former school janitor gave up that full-time occupation when he was informed that, if he remained employed outside the fishery, he would have to give up his salmon and lobster licences. As he explained,

I'd rather be fishin' than scrubbin'. (Respondent no. 018)

The only negative views of the fishery as an occupation came from part-timers. One, though claiming to have no other work, declared,

I don't like anything about it, but you got to. I suppose I likes to be at it in a way, but it's an awful racket. (Respondent no. 013)

Most of the King's Cove fishermen, like those in Charleston, were engaged 'at the squids.' Their wives often helped to dry the squid, and several held fishing licences themselves in order to qualify for unemployment-insurance payments. Again, as in Charleston, some people resented or even condemned such practices:

People here with good jobs, they're not even part-time fishermen. They are able to go out and catch squid and sell 'em in their wife's name and she gets the unemployment [insurance]. I don't know how you can have a fishing licence and not go out in a boat at all and still get UI stamps. (Respondent no. 016)

You might cut some of these fellows out who have a $25 punt and want the same benefits. I would cut out the women who are never on the water. (Respondent no. 018)

One woman who was 'at the squids,' however, justified her efforts by noting that the income she earned enabled several of her children to go to technical college and university:

I goes at a few squid to help my kids through school. (Respondent no. 021)

Although squidjigging was important, the main focus of the fishery in King's Cove was in fact the lobster fishery. When asked what they would do if they couldn't get a lobster licence, all those who held such licences promptly declared that they would have to give up fishing. Some even suggested that they would end up on welfare.

I'd go on welfare! Mark that down loud and clear. That'd be it for me. (Respondent no. 013)

I'd look for another job – and, as a last resort, welfare. (Respondent no. 017)

If they take the lobster licence away from me it'd be a lost cause. (Respondent no. 018)

'It was an unwritten law.' (Respondent no. 022)

According to *all* those we interviewed, King's Cove has no formal system for allocating fishing berths:

The way it is here, you puts your net where nobody has got his. (Respondent no. 016)

We fellows, wherever you puts one [i.e., a net], that's it. (Respondent no. 013)

As we have said before, the presence of some sort of system of community regulation is critically important. If there is no such system in place, the possibility of unbridled competition remains high and the tragedy of the commons is a likely outcome. By contrast, the presence of such a system supports the argument that the inshore fishery is a form of community-regulated property, rather than the open-access resource posited by the neoclassical model.

It is significant that, while they denied the existence of any system of community regulation, several respondents also noted that there was no need for such a system. They argued that community members had always respected the rights of others by putting their nets and traps in the same place every year and by avoiding locations where other people had customarily placed their gear. In other words, the system of 'traditional occupation' was so well established that none of our respondents thought of it as a community based 'system' at all. It was simply the way things were. Thus, the same two people who were quoted above as saying there was no such system went on to state,

Nobody won't go and put their nets where I been putting my nets every year. If someone did, that's all you could do. But that's never happened yet. (Respondent no. 016)

I puts mine in the same place each year. (Respondent no. 013)

In a similar vein, another added,

Here we don't have no draw [i.e., lottery for berths]. Say I had a berth last year, usually I can get my net back there this year. Unless some fellow wants to act different and take your berth. But there's an awful lot of fireworks over that. (Respondent no. 018)

The latter fisherman went on to describe how the system actually operates in practice. As he explained it, a few days before the start of fishing season, he would put a mooring in a location where he intended to set a net, as a way of signalling that intention to others. He admitted, however, that such a mooring would not 'legally' hold a berth:

A day or so before the season, you puts your mooring there. Though really, a mooring don't hold a berth. There should be some other way. (Respondent no. 018)

Although this customary allocation of berths was corroborated by several other people, there was also considerable evidence to suggest that the system was not without its problems. An older fisherman suggested that the younger fishermen no longer respected traditional practices in this regard:

Elderly people always respected where another fella put his gear. It was an unwritten rule. But now, we got a fresh batch who observes no rules or regulations. (Respondent no. 022)

The same respondent went on to complain about part-time fishermen who waited until they saw someone else catching salmon or cod, then surrounded his nets or traps with their own. Although it might be argued that, technically, they were respecting his right to set his gear in a particular location, they were also virtually ensuring that few fish ever reached his gear.

Whenever they see a man getting a few salmon or if he is handling fish [i.e., codfish], they slap gill nets around him and that finishes him. I've been poisoned with it the past two or three years. People turned their noses up at those who are fishing, and then they make the rush out in a boat with a government subsidy. (Respondent no. 022)

Another fisherman told a similar story, about trying to haul his net, only to get his motor caught in the nets of other fishermen, who had set theirs so close to his as to make his task impossible:

I had to cut a fella's gill net off Saturday. It was caught into my motor. There was no need for it. I was there long before them and they all came down and plastered their nets down on top ... It seem like every year it is getting worse. I got a clean cut for him, but I ruined mine. He set on top of me. (Respondent no. 017)

It must be noted, however, that this respondent, like several others, blamed the problem not on fishermen from his own community, but on those from other communities. Indeed, this became a common theme throughout our interviews with King's Cove residents. Lack of respect for King's Cove's traditional gear locations was almost invariably blamed on those from other communities. Similarly, our respondents complained about outsiders' overcrowding the lobster beds traditionally used by King's Cove residents:

They set their pots right on top of you [i.e., your pots]. They smother you with pots. Nothing you can do except have a scattered argument with 'em. (Respondent no. 018)

Several respondents also complained about fishermen from other communities who would leave their boats in King's Cove and reportedly occupy most of the space at the wharf that was available for cutting up fish and tying up boats:

This is the worst place for people from elsewhere. They comes in here and takes over the wharf and they got no regard for the people from here. They just leaves their boats tied on. (Respondent no. 018)

A lot of other people from other communities come here and tie on their boats and take up our wharf space and we only has room here for our own boats. (Respondent no. 023)

Such people were undoubtedly taking advantage of the new breakwater and the new government wharf. Some were from nearby communities. Others were from 'up the bay' – from places like Charleston – where the fishing was poor. They were attempting to maximize their fishing effort by fishing out of a more favourable location. Although their efforts would likely be praised by fishery officials and cited as a justification for government expenditure on fishery infrastructure, their presence did not sit well with King's Cove residents. Still, at least one respondent was philosophical about the intrusions, noting that he had done the same in previous years:

Let 'em come. You're only going to make a dollar whether twenty-five more comes or not. Probably squidjigging you'll end up somewhere else. Two years ago I spent two months in Trouty jigging when there was no squid here. (Respondent no. 019)

Despite the possible legitimacy of the reasons for them, these intrusions by fishermen from other communities nevertheless cast some doubt on the effectiveness of the practice of 'traditional occupation' as a means of regulating the commons.

'They should be fishing for want.' (Respondent no. 013)

The licensing of fishermen is the state's chosen 'solution' to the potential problem of unfettered competition stemming from the common-property nature of the fishery. Yet, as we saw in our discussion of Charleston, that solution can create other sources of conflict and competition. The two main dimensions of fishery licensing that can be sources of conflict are (1) the distinction between full-time and part-time licence holders and (2) restrictions on the number of protected-species licences that are issued. In King's Cove, as in Charleston, both dimensions gave rise to conflict within the community.

Some of the fishermen we talked to expressed a general concern that simply too many part-timers were allowed to fish in the community. Indeed, in King's Cove, the conflict was based not so much on the distinction between full- and part-time fishermen *per se*, as on the fact that part-timers were allowed to fish commercially at all:

I think [licensing] should be made a lot more stringent. There's too many people in the fishery that is the problem. Half of them should be cut out of it. (Respondent no. 017)

Half in Newfoundland fishery shouldn't be at it at all. If they are fishing, then they should be fishing for want. (Respondent no. 013)

The brunt of the concern was the role of people who were referred to locally as 'moonlighters.' These were people who held generally well-paying, full-time jobs and who also engaged in the fishery:

We go out past moonlighters in the morning to haul our gill nets. It comes on windy, and we can't haul one and we, OK, we'll go in and jig a squid. And what happens, the part-timers got that all blocked at the plant and we can't sell a one (Respondent no. 018)

Some protested against moonlighters in general terms:

Unskilled workers who don't get a job, well he got no other choice but to go fishing. He got to live too. What are you going to do with him? It's fellows with permanent jobs shouldn't have [a part-time licence]. Here, they all jump on the bandwagon once the squid start, which is not very fair to us. (Respondent no. 018)

See a fellow classified as a part-timer could be a doctor, or a fellow working with the railroad, can still get a cheque from CN. But I get my only cheque from fishing. (Respondent no. 018)

I don't see any difference in full-time and part-time fishermen. He gets the same price for his fish as we do. (Respondent no. 023)

In King's Cove, the primary villains with respect to moonlighting were the schoolteachers, of whom there were many in the community because it was the educational centre of the region. Most of the male teachers were the sons of fishermen and had grown up in small fishing villages. They had consequently learned the basic skills of fishing in their youth. Because their jobs as schoolteachers usually allowed them their summers off, they were in an excellent position to earn a second income from fishing. There can be no doubt that their presence was strongly resented by the traditional fishermen:

There shouldn't be schoolteachers having a licence and selling fish. (Respondent no. 013)

Everybody be at it – schoolteachers and everybody ... A part-time fisherman can be anything – a schoolteacher or anything at all. Could have the biggest kind of a job, sure. (Respondent no. 016)

You can't go teaching without qualifications. Why should the fishery be the only thing you can do without qualifications? (Respondent no. 018)

Judging by their comments, the basis of these fishermen's concerns appears to be that the part-time fishermen are catching fish that rightfully belong to them, that is, to full-time fishermen who have no other source of income. But the vehemence with which King's Cove fishermen expressed their complaints suggested that there was more to the problem. Even by urban Newfoundland standards, teachers earn an excellent income. Given the low cost of housing and the generally low incomes in rural Newfoundland, teachers' incomes are considered by many to be astronomical. That these people should also have the right to earn a second income from fishing is seen as terribly unfair:

A full-timer depends on it for a living. Part-timers – some of 'em plays at the fisheries. (Respondent no. 017)

I don't mind anyone having a boat to get a few fish for themselves. But I don't believe in 'em scabbing. (Respondent no. 013)

The complaints about teachers' and other part-timers' involvement in the fishery were particularly bitter when they pertained to part-timers who also held protected-species licences. Again, some of this anger was expressed in general terms:

A full-time fisherman should get all limited-entry licences, whereas a part-time fisherman should only get the cod licence. There are part-timers with other licences – shouldn't be. (Respondent no. 017)

If a man got a job, they shouldn't have a full-time licence, nor a salmon or lobster licence. (Respondent no. 023)

Others continued to express their anger specifically, as an attack on teachers:

There is people who goes out fishing, like schoolteachers who get [protected-species] licences. A lot got 'em and don't use it. A lot of people wants salmon and lobster licences and they can't get 'em. (Respondent no. 023)

The greatest hostility, however, was reserved for old-age pensioners who retained lobster and salmon licences. So great was this resentment that one suspected it had to be fuelled by some deeper, intergenerational, conflict.

Pensioners shouldn't have a licence. If they need it, give up the pension and go fishing. (Respondent no. 017)

There's more young fellows can't ge a licence and yet here [in King's Cove] there's old-age pensioners and schoolteachers with licences. I don't agree with that. (Respondent no. 016)

There is men here getting their old-age pension and they can get lobster and salmon and the young fellows can't get nothing. (Respondent no. 021)

There's a lot of people shouldn't have 'em [i.e., protected-species licences]. People have had 'em for years and years and won't give 'em up – old-age pensioners. (Respondent no. 019)

Whether such hostilities would exist even in the absence of a licensing system is, of course, an unanswerable question. There can be no doubt, however, that the licensing system as currently constituted serves to divide King's Cove. Indeed, if King's Cove and Charleston are any indication, conflict within communities arises not in relation to local community control of resources, but in relation to the implementation of the state's attempts to control those resources through licensing regulations.

'You see, the fishery has always been a game that, when
other employment fails, I'm going fishing.'
(Respondent no. 022)

In comparison with Charleston residents, the people we talked to in King's Cove seemed less obsessed with devising strategies to ensure receipt of unemployment-insurance benefits through the winter. This seems attributable, in large part, to the more stable fishery at King's Cove. As the quotation from a King's Cove fisherman that opens this chapter says, in this community 'we got no trouble getting clear of our fish and getting a good price for them too.' King's Cove fisherman certainly referred to the fishery as 'a gamble' and a 'game':

Fishing is only a gamble. (Respondent no. 019)

You see, the fishery has always been a game that, when other employment fails, I'm going fishing. (Respondent no. 022)

However, these fishermen were not nearly as caught up in the 'calculus' of hunting an income from fishing as were the residents of Charleston.

Although King's Cove fishermen may not have been dominated by the struggle to win the game (or even to survive it), they were not unconcerned about those whom they perceived to be playing it unfairly, cheating the system or benefiting from government assistance programs for which they did not qualify. Some of the complaints pertained to activities that, if true, probably bordered on the illegal. Information of this sort is always problematic for a researcher. First, it is only hearsay, and its truth and accuracy are open to question. Second, to present it in detail would be to risk revealing the identity of both the informant and the individuals involved. Consequently, such information will be presented here only in very general terms.

As in Charleston, we heard complaints about people who were just 'fishing for stamps' – that is, people who quit fishing once they had their required number of unemployment-insurance stamps. We also heard complaints about women who dried squid caught by their husbands and then claimed unemployment insurance for it. We were told of instances of the latter where the husbands were fully employed at other occupations. Among such reports were stories about people who received land stamps rather than fishery stamps for squid because 'they had friends in the right places.' We also heard of individuals who were collecting unemployment insurance full time from previous land-based work while engaging in squidjigging and other types of fishing. It was reported that the fish caught by such people was being sold by other relatives, who then collected the unemployment-insurance stamps for them. We listened to complaints about fishermen who had more lobster pots in the water than their licences permitted, and heard stories of people 'from other communities' who used boiling water to wash the spawn off fertile lobsters to make them acceptable for sale. Finally, some of our respondents attacked the government programs for awarding subsidies for the purchase of longliners, because they felt that the people who had received them were not capable fishermen. The last word on the subject is best reserved for the fisherman who complained that 'there is always someone covering up for one and uncovering for another.' (Respondent no. 013).

'I haven't got a union card, but I thinks I'm into him.'
(Respondent no. 016)

As noted earlier, the major external agencies outside the state that affect all inshore fishermen are the fishermen's union and the various fish buyers. If it is primarily the government that determines who has the right to fish commercially, it is primarily the unions and the fish buyers who, together, determine how much income can be obtained from the fish that is caught.

There were few complaints about the fish buyers. Most King's Cove fishermen had a wide array of fish plants and buyers to whom they could sell their fish. Some sold fish through agents in nearby Plate

Cove and Dunterra. Now, however, most sold their fish to the plant in Charleston. This situation had come about when the Charleston plant guaranteed that, if King's Cove fishermen supplied it with fish on a regular basis throughout the rest of the fishing season, it would take their fish during a glut.

With respect to the union, some of the part-time fishermen did not know whether they were members and some other fishermen said they had joined because they had no other choice:

I haven't got a union card, but I thinks I'm into him. I'm paying into it every week when I ship fish. (Respondent no. 016)

I had to join. They takes your dues out. We had no other choice but to join. (Respondent no. 023)

There were also complaints that the union hadn't 'done us any good' (Respondent no. 016) and that it had given too many benefits to plant workers and trawler fishermen at the expense of the inshore fishermen:

I was more less forced into it. I'd like to get out of it. They're giving everything to the plant workers and trawlers, and we're getting nothing. (Respondent no. 017)

However, such comments are generally to be expected from workers who are not involved in the union movement and who are members only out of necessity. Indeed, the most interesting aspect of union involvement in King's Cove was that some fishermen expressed considerable support for the union, much of it in appreciation of the higher prices that the union had been able to obtain:

It's got better prices for fish. (Respondent no. 020)

It's done the fishermen some good ... No big lot, but some. (Respondent no. 013)

One person also emphasized other personal benefits that came with union membership:

For one thing, we got a drug plan. We gets 85 per cent of our drugs paid for now – and [those of] our wives and children. We're covered under workmen's compensation and the union brought that about. They're keeping up the price of fish. And, some of the ways the fishermen are treated around the plant – they got a say in that. They come up with an accident insurance now for anywhere in the world. (Respondent no. 018)

The main complaint about the union centred on its involvement in the regional Fisheries Licensing Appeals Board. This is the body that hears appeals from fishermen who are unhappy with their part-time classification. Members of the board are appointed by the government, on the recommendation of the union. Some respondents were vocal in their complaint that at least some of those appointed were not among the most experienced fishermen:

I was surprised to hear that [a resident from the area] was on the appeals committee. I wouldn't say that he was experienced enough. (Respondent no. 015)

I don't know how [a particular fisherman] got picked, I can tell you ... When I heard [he] was on it, I thought it was too foolish to enquire about. He got on because he had no lobster licence or salmon licence for himself and he thought if he got on it he'd get one. So he got on it and he got the salmon licence. (Respondent no. 018)

Such comments are in line with some of the previously cited complaints by King's Cove fishermen. Although they may view their community positively as a place to fish, they are critical about the ways in which some of their neighbours use, and perhaps benefit from, the regulatory system. Certainly, it would appear that all such actions are the subject of open comment in King's Cove.

'If I was a young man, I wouldn't hang up here.'
(Respondent no. 013)

In a country in which people are free to move around, it is to be expected that the majority will like where they live. King's Cove is no exception. The response to almost all the questions in the category 'Evaluation of Community' in Table 5.2 was extremely positive.

When it came to community services, schooling and medical services were given a very positive evaluation:

TABLE 5.2
King's Cove respondents' views of their community and its future

	Yes	No	Uncertain
Evaluation of community			
Like living here	9	1	0
Good place to raise children	9	0	0
People here generally hard-working	8	1	1
Leaders are good, capable people	6	1	3
People here care about the place	8	1	1
Evaluation of community services			
Good schools and teachers	9	0	1
Good medical care	8	1	1
Enough to do in spare time	0	6	0
People here do without a lot	5	5	0
Evaluation of employment opportunities			
Good place to find work	0	9	0
Fishery has a future here	5	1	2
Fish plant here has a future	2	2	3
Inshore fishery good for young people	2	6	1
Evaluation of community's future			
Children should settle here	0	6	2
Community has a good future	5	4	1

Note: There were 10 respondents. Where responses for any given item do not total 10, the balance represents abstentions.

We've got two of the best schools in Newfoundland. (Respondent no. 013)

We got a good school here, boy. The teachers are first-rate. (Respondent no. 016)

However, there was a mixed reaction to the question of whether people had to do without a lot, and a strong belief that there was not 'enough to do' in the community.

Respondents were united in their belief that King's Cove was not a good place to get work, and divided on whether the community and its fishery had a future. It is of particular interest that very few respondents regarded the inshore fishery as a desirable occupation for a young person. Not one wished his or her own children to settle in the community after finishing school:

When they finish school and university, they got out of King's Cove. There's nothing here for 'em. If I was a young man, I wouldn't hang up here. (Respondent no. 013)

I'd like to see them stay, but they're going to have to get out of it. There's nothing here to do. (Respondent no. 020)

Because of the schools, King's Cove is considered by its residents to be a good place to raise children. But, unlike the residents of Charleston, those of King's Cove generally did not want their children to remain in the community: The road from the school really does lead out of town.

Conclusion

This book is about the intersection of state policy and community life in fishing communities. State policy is built on the premise that in the absence of externally imposed regulatory measures, fishermen will operate essentially in a 'state of nature' characterized by conflict and unbridled competition. The two communities discussed in this chapter offer evidence that, contrary to this belief, local community involvement in fishery regulation does exist. Indeed, the challenge for fishermen in communities such as these is how to integrate state regulatory efforts with their own, already-existing system of regulation. Although it is clear from the responses of our interviewees – that is, from their *descriptions* of their fishing activities – that both Charleston and King's Cove have traditional ways of regulating the location of fishing berths, not one respondent from either community answered positively when asked directly whether the community had 'some way of controlling where nets and traps are set in the water.' If it does nothing else, our study thereby gives evidence of the value of qualitative over quantitative analysis in this area of inquiry. Both of the communities practised a system that we have called 'traditional occupation,' whereby the right to place one's net in a specific location is based on previous occupation of that location. None of our respondents thought of this as a system of community regulation, because they accepted it as simply a 'natural' condition of fishing in their communities.

In this chapter we have considered many of the issues that were highlighted in chapter 1. For example, we provided evidence of the difficulty of distinguishing between full-time and part-time fishermen, and outlined the implications of government attempts to do so for the employment activities of both types of inshore fishermen. We also noted the gamelike nature of earning a living from fishery and the extent to which fishermen engage in a sort of calculus to determine the best ways of earning an income. Our analysis led us to consider the implications of the assertion that fishermen are hunters involved in a twofold hunting process: (1) the hunt for fish and (2) the hunt for the income to be derived from the sale of fish. We have seen that it is primarily the licensing system that sets the conditions under which the hunt for fish takes place, and that it is primarily the unemployment-insurance system that influences the nature of the hunt for income from the sale of fish. Both are state policies that effectively shape and constrain the nature of work and community life in Newfoundland inshore fishing communities.

We have also seen proof of our claim that fishermen fish from communities as well as from boats. Not only is the practice of fishing in Charleston and King's Cove constrained by each community's particular pattern of regulation, but the fishermen of the two communities differ quite considerably in the ways in which they carry out their fishing activities.

Most importantly, we have found ample evidence to support our claims about the types of property relations that may exist in resource-based rural communities. Such evidence was revealed in a variety of contexts, from the strong belief among fishermen that their protected-species licences should be private rather than state-controlled property to the way in which fishermen from nearby communities recognized Charleston fishermen's rights to engage in perimeter defence of their local fishing grounds. In short, the evidence we found attested to community regulation. Further evidence that such traditional property rights continue to exist will become apparent in the next chapter, in which we discuss two more 'small worlds.'

6 Community Control and Conflict

Regulating Competition in Two Intermediate-Sized Fishing Communities

This is one of the richest communities on the island.

A Grates Cove fisherman
(Respondent no. 077)

This is the best place on the island for fish. They hangs around here until the end of October.

A Fermeuse fisherman
(Respondent no. 098)

Grates Cove and Fermeuse, the two communities discussed in this chapter, differ from each other considerably and have quite different types of fisheries. They provide further evidence of the community character of the inshore fishery and of the distinctiveness of each community's fishing organization. As in chapter 5, the analysis of the two communities will be undertaken in the context of the issues outlined in chapter 1. We will show that the definition of what constitutes a fisherman and the distinction between part-time and full-time fishermen are important issues in Grates Cove and Fermeuse, just as they were in Charleston and King's Cove. Similarly, the fishermen's approach to earning a living from the fishery in these communities also has a gamelike quality of 'calculating' optional strategies. In these communities, as in those described earlier, state regulations are important in defining the terms and conditions of the 'game.' As a result, in each of these communities there is ample evidence of an established system of rules and regulations governing the manner in which work in the fishery is to be conducted.

In Grates Cove and Fermeuse, however, such rules do not take the implicit, customary form that they do in Charleston and King's Cove. Rather, they are an explicit set of regulations to which, significantly, the residents of both communities refer as 'community laws.' In Grates Cove these rules originate in local custom; in Fermeuse they appear to be a combination of local custom and state intervention. So pervasive are those community laws that 'community control' becomes a pervasive issue throughout this chapter.

A second characteristic feature of the fishery in both these communities is gear conflict, arising out of fishermen's use of different 'means of production.' In Grates Cove gear conflict develops as a result of competition between inshore boats and longliners, which use different types of technology to catch fish. Specifically, the smaller boats use traps and the longliners make extensive use of gill nets. In Fermeuse gear conflict occurs not between different types of boats, but between similar boats that use different technologies – specifically, trawls, gill nets, and handlines. Here, the conflict is also rooted in a long-standing dispute between Fermeuse and a rival community for the same fishing grounds. Much of Fermeuse's social organization of the fishery is built around these various 'community conflicts.'

Many political economists distinguish between the *forces* of production – that is, the physical and technological aspects of production – and the *relations* of production – that is, its social dimensions. However, in the context of the fishery, and to better articulate the nature of the conflicts occurring in inshore fishing communities, we have found it useful to employ a further distinction within the physical aspects of production, namely, between means of production and technology of production. We use the term *means of production* to refer to types of boats, and the term *technology of production* to refer to the type of equipment used in the boats. The boats used in the fishery are, we believe, analogous to the factories that are the means of production in heavy industry. And just as the equipment in different factories that produce the same product can be quite different in terms of technological sophistication and the type of work organization required, so too can the equipment on different fishing boats be different in terms of technological sophistication and the nature of the social relations involved in its operation. As we will see, this set of distinctions is at the core of the gear conflicts and, hence, of the social relations that develop because of them in the inshore fishery.

Finally, it is important to note that the following analysis permits us to see more clearly the distinctive regulatory roles played by the state and the community. Both the state and the two communities under discussion have engendered extensive regulations to control access to the fishing grounds. But whereas the state makes comparatively little effort to regulate the technology that may be used there, this aspect constitutes a predominant concern of community-based common-property regulation.

GRATES COVE

Grates Cove is located near the tip of the peninsula that separates Trinity and Conception bays. The houses huddle in hollows along the top of a 150-metre windswept cliff. No trees grow here, for they could not withstand the almost constantly battering winds. Even on a warm and sunny summer's day, one has to lean into the wind. In bad weather, anyone going too close to the cliff edge runs the risk of being blown over and hurled against the rocks below. Far below, down a steep lane that affords the only access to the stone beach, lies the small community wharf, battered and slightly broken up by last winter's storms. The trapboats toss and veer at their moorings in the open harbour – it is too dangerous to leave them tied to the wharf for long. Even the smaller runabouts that are used to ferry back and forth to the larger boats are usually hauled up a slipway clear of the water when they are not in use. Conversations with Grates Cove fishermen often include some mention of the harshness of the location or of the wind and the sea conditions that are produced by it. After all, these are the environmental conditions that shape the character of work here.

It's a hard place to fish. You got to be like a cat to fish out of a place like this ... This is a rough area. (Respondent no. 083)

You come here in the early part of October, and you got no business being out there [i.e., on the water]. We got no harbour here. (Respondent no. 089)

You take September and October and you might get out [i.e., fishing] only a week. We got bad weather here. (Respondent no. 095)

There's rough conditions here. You're not able to land a big load of fish when it is too rough. (Respondent no. 089)

There is only one obvious reason for people to continue to live and fish in as inhospitable a location as this: The fishery must be 'awfully good.' Throughout most of Newfoundland, fishermen tend to be reticent about describing their catches in positive terms, as if to do so would be to tempt fate. Not so in Grates Cove; here, fishermen talk with pride about the success of their fishing:

This is one of the richest communities on the island. (Respondent no. 077)

A conversation with one of the fish buyers who regularly collects fish from the community produces data that justify such sentiments:

I haven't seen [the fishery] fail in the four years I was buying here. The first year I bought two million pounds from ten to twelve crews. And this year me and the fella from [a rival firm] bought one a half million pounds between us.

Grates Cove was known throughout its long history – and is still known to its current inhabitants – as 'The Grates.' John Guy, when he sailed into Conception Bay in 1610 to establish the first colony in Newfoundland, wrote of passing 'The Grates' at the entrance to the bay (Prowse 1895, 133). Although the census of 1675–7 did not record any inhabitants in the community, nearby Old Perlican, to the west, had more than fifteen families, and Bay de Verde, immediately to the east, had approximately five (Handcock 1977, 17). It is therefore likely that The Grates was inhabited soon afterwards. Certainly, the family names now to be found in the community have a long recorded history. Seary (1977) records a John Snelgrove and a John Blunden as occupiers of fishing rooms in Grates Cove in the winter of 1800–1801, as well as a Samuel Broderick in 1806. Records also list a Mary Doyle in 1808, a 'planter' (that is, a settler) named George Lambert in 1825, a Mary Avery in 1830, and a William Meadus in the 1830s. Despite this long history, however, not one person interviewed in Grates Cove mentioned the origins or duration of the community. Unlike the residents of King's Cove, these fishermen were concerned not with the past, but with the present and future of their community and its fishery.

The 1986 census shows 275 residents in Grates Cove, which is only 20 more than in King's Cove and 33 more than in Charleston. However, there were 72 licensed fishermen living in the community, of whom fully 57 held full-time licences and only 15 held part-time licences. None of these licence holders was a woman. In addition, there were 21 salmon licences, 3 herring licences, and 1 lobster licence. These were held by 23 different people. The comparison with Charleston and King's Cove is striking: Those communities had only 13 and 10 full-time licence holders, respectively. Clearly, in contrast with other communities of similar size, Grates Cove is a major centre of fishing activity.

We interviewed 22 of the 57 full-time fishermen in Grates Cove. Because of time constraints and financial limitations, because part-time fishermen constituted only a relatively small proportion of the fishermen in Grates Cove, and because our interviews in other communities had indicated that part-time fishermen were generally not well informed about norms and practices in the fishery, we decided to interview only full-time licence holders in this community. Of the 22 fishermen we interviewed, 11 held salmon licences, including 1 who also had a lobster licence.

*'The father and sons, and the two sons together. That
is how crews is set up here.'
(Respondent no. 079)*

The organization of the fishery in Grates Cove is more complex than that in Charleston or King's Cove, and requires some preliminary explanation. Most fishermen in both of the latter communities prosecuted the fishery in relatively small, open boats. In Grates Cove, unless one is only, literally, a 'fair-weather fisherman,' the prevailing conditions of sea and weather require something more substantial. Of the people we interviewed, only three fished from small boats with outboard engines: One was an elderly, semi-retired person who had 'fished his lifetime' in bigger boats and was now simply 'passing time,' and two were younger men whose involvement in the fishery was somewhat marginal.

The 'traditional fishery' in The Grates has centred around the trapboat and the cod trap. The trapboat is a substantial boat with a deep keel, wide at the midships, and with an inboard engine. Fifteen of the nineteen people with more-substantial boats whom we interviewed fished from trapboats that ranged from 24 feet (7.3 metres) to 31 feet (9.5 metres) in length and were powered by diesel engines of 20–35 horsepower. The other four were longliner fishermen. The longliner is a deep-water boat with high bow and sides, deep keel, and square stern. The boats of the four longliner fishermen from The Grates ranged from 38 feet (11.6 metres) to 51 feet (15.6 metres) in length. Because of the poor harbour and the lack of protection from storms in Grates Cove, these longliners actually operate out of Old Perlican, about 24 kilometres away by road. However, the men who work on them continue to live in Grates Cove and are therefore able to participate in community draws for both cod-trap and salmon-net berths near that community. (They would not be allowed to participate in similar draws in Old Perlican.) Thus, despite fishing from and returning to a port other than Grates Cove, they continue to be regarded by all parties as Grates Cove fishermen.

In part because of the significance accorded in the existing literature to the differences in social organization and property relations between trapboat and longliner crews, we asked our Grates Cove respondents a number of additional questions about crew structure, boat ownership, and income division. Like previous researchers, we found that trapboat crews are almost invariably based on kinship. Two patterns of owner-ship and control can be observed. If a father and sons form a crew and the father is still active, then the father will usually be the skipper and boat owner, and the sons will be sharemen. Exceptions to this rule occurred only when the sons had been fishing with the father for more than a decade. By contrast, if brothers are involved together in the absence of the father, it is likely that they will remain equals, as co-owners and 'co-skippers' of the boat. Additional younger relatives, such as sons, nephews, or grandsons, may be part of such sibling-based crews, but only as sharemen. Among trapboat crews, the owner or owners of the boat and equipment take the largest percentage of the income from the catch. Thus, in one trapboat operated by a father and his two sons, the father/skipper received 50 per cent of the income from the catch, while each son received 25 per cent. The same distribu-tion was found in a trapboat in which an experienced fisherman and boat owner fished with his less-experienced brother-in-law and another, more distant, relative. In another crew, two brothers and the son of one of them had fished together for more than fifteen years. Because they had worked together for so long and because all three now had equal ownership stakes in the equipment they were using, they all got equal shares of the income.

Longliner crews are also formed primarily along kinship lines, though there may be exceptions. One boat for which we have information was manned by a father and his four sons. As in the case of trapboats, the father was the 'skipper,' reflecting the fact that the ownership (and the financing) of the boat was recorded solely in his name. In this relation-ship, however, the boat was treated as a distinct 'company' and was awarded 40 per cent of the income derived from fishing, with the crew members and the skipper each receiving a 12 per cent share. Exactly the same arrangement was found on a longliner operated by a father/skipper, his three sons, and his son-in-law. In a third case, the skipper was unrelated to any of his three-man crew. In this instance, the boat still received a 40 per cent share, while the skipper and each crew member received 15 per cent.

As we noted in chapter 2, it is the sort of financial relationship just described that caused Fairley (1985) to view longliners as capitalistic operations and prompted Sinclair (1985, 142) to consider them an example of 'dependent petty capitalism.' It is our opinion, however, that too much has been made of the differences in financial arrangements between trapboats and longliners and that too little attention has been paid to conflicts relating to differences in the technologies of production used on the two types of boats. As our Grates Cove data show, trapboats often fail to exhibit the equality of relations that tends to be associated with them. At the same time, because of the level of debt involved, few longliner skippers make a profit from their boat investment. Thus, although longliner skippers may, theoretically, be in a position to 'capitalize' on their supposed ownership and control of the means of production (the boat), most of them consistently fail to do so.[1] Consequently, it seems academic to debate whether or not they can be called capitalists.

In contrast, as we shall see later in the chapter, the type of technology used in production is of tremendous social significance. In Grates Cove the technology of production that has traditionally been employed is the cod trap. As we explained earlier, the cod trap is a net box with a long net 'leader' that is submerged in the water in locations fish are known to frequent. The twine in the leader is large enough for fish to see it and to attempt to avoid it. When they encounter the leader, their instinct is to swim alongside it towards deeper water. As they do so, however, they pass through a narrow opening into the trap and are unable to find their way back out. They remain alive in the trap until the fishermen gather it up and scoop them out. The season for using cod traps is usually short, because they require large schools of fish swimming relatively close to the surface, and this occurs only when the cod are chasing schools of capelin, which come to spawn on the beaches of Newfoundland every year.

Grates Cove is largely a trapboat community, a fact on which it prides itself. As we shall see later, most of the trapboat crews have two traps in the water[2] and derive about half to two-thirds of their annual fishing income from them in a short, six-to-eight-week period.

Most have two and a lot of guys have three [i.e., traps]. And if a guy starts out with one cod trap, the next year he'll have two to cover more territory. With [only] one trap, all your eggs are in one basket. (Respondent no. 077)

When the trap season is over in early August, the trapboat crews switch to handlining. A handline and jigger consist of a weighted hook attached to approximately six to fifteen metres of line. The jigger is lowered close to the bottom, where its up-and-down movement, called 'jigging,' attracts the attention of the fish. As the fish come close to investigate, they are hooked and then quickly hauled up into the boat. Although it is a very labour-intensive method, handlining can produce sizeable catches in areas where groundfish are plentiful. Grates Cove is situated near one such area.

Trap fishermen are perpetually threatened by the gill-net technology that is the primary means by which the longliners harvest fish. Gill nets are made of a fine mesh on which fish catch their gills when they swim into them. Unable to breathe properly, the fish drown and hang off the net until they are removed. Gill nets can be anywhere from fifteen to thirty metres long and are set either in open water or at the entrance of bays and inlets. A large 'fleet' of overlapping gill nets placed at the mouth of a bay or near prime trap berths can make it virtually impossible for fish to reach the traps. The four longliners represented in our sample had 165, 175, 200, and 300 such nets, respectively. Only a portion of the nets were hauled every day, meaning that some fish might remain in the nets several days before being collected. This was particularly the case when bad weather and heavy seas made fishing impossible.

The full implications of the aforementioned aspect of gill-net use will become apparent later in the chapter. Put simply, however, gill nets have the potential to destroy the traditional trapboat fishery. Consequently, trapboat fishermen are vehement in their condemnation of gill-net fishing:

The longliners got everything overfished around here. Gill nets was the greatest disgrace ever done to this country. I say they should be done away with altogether. It is a menace. (Respondent no. 087)

If they can't get clear of the gill nets altogether then get an inch mesh. That way all you get is very big cod. Or set a certain time of the year for setting gill nets in order to let the fish in. If we are all out there with gill nets, we are starving the fellas in small boats. In the spring of the year put out eight-inch mesh, and after a few weeks put out five-inch mesh, and give the cod a chance to come in. (Respondent no. 085)

Before the gill nets come, the fish came from everywhere. But since the gill nets and longliners started, every year is getting worse and worse. 'Cause every year they fences off Trinity Bay with those gill nets. (Respondent no. 082)

In sum, despite the considerable amount of academic discourse about class differences arising out of the nature of ownership and control of the means of production in the inshore fishery, it is the argument here that such differences are of limited relevance in the day-to-day activities of inshore fishermen. There are, however, two fundamental issues that can indeed divide inshore fishermen. The first is gear conflict – that is, the conflict that arises as a result of competition between different technologies of production. The second involves the distinction between full- and part-time fishermen, and the appropriate allocation of protected-species licences. Both of these are social issues, in that they affect the nature of work and community social relations. And, as we shall see, whereas the fishermen of Grates Cove have exercised some level of community control over the problem of gear conflict, the issue of licensing remains largely beyond community control.

'I'm a Newfoundlander and the sea is in my blood.'
(Respondent no. 082)

Grates Cove fishermen like their work. The fact that none had been employed outside the fishery during the preceding year notwithstanding, one of the most significant things about Grates Cove fishermen is that many of them returned 'home' to fish after many years of employment in other occupations and in other communities. One man had managed a small fish-processing plant in another community; another had spent more than a decade selling real estate in California; one was a former bartender; two men had spent several years at factory work in Toronto; one was a former brewery worker; one had worked five years as a security guard in St John's; and two had worked in fish plants in other communities. All these men had returned to Grates Cove because 'the sea was in their blood,' and they wanted to fish.

Only two of the people we interviewed indicated any dislike for the fishery, and they objected mostly to the long hours. The others were almost ecstatic in the love they expressed for both their work and the way of life that went with it:

It's the hardest kind of work but you don't seem to notice it. It's so interesting. (Respondent no. 089)

I was reared up [i.e., raised] at it and I loves the job. There's nothing I don't like. (Respondent no. 097)

The work. I enjoy it. There's nothing really I don't like. (Respondent no. 091)

When probed about what it was they liked so much, several said it was important to them to be their own boss:

You're your own boss and it's up to you to do it. I love it. It's the only way to fly. (Respondent no. 072)

Fishing – it's up to you to make it. You're the boss. (Respondent no. 077)

One thing that was unique to Grates Cove, however, was that some respondents (eight of twenty-two) claimed they liked fishing primarily because of the money there was to be made in that location:

There's good money here when you get the fish. (Respondent no. 076)

The fishermen we interviewed in Grates Cove were different from those in other communities in other ways as well. They were significantly younger than those we interviewed in King's Cove and significantly better educated than those we interviewed in either Charleston or King's Cove. In Grates Cove, 54.5 per cent of our interviewees were under age forty, and 36.3 were under age thirty. In comparison, in King's Cove, only 27.3 per cent were under forty, and 63.7 per cent were over fifty. In Grates Cove, fully 72.8 per cent had at least a grade-nine education, and 36.4 per cent had completed high school. None of our Charleston interviewees had completed high school, and only one in King's Cove had done so. It is likely that Grates Cove has one of the youngest and best-educated complements of fishermen of any Newfoundland community. This ability to hold its well-educated young and to attract back some of those who do leave is significant to Grates Cove residents. To them, it is proof of the vitality of their community and its fishery:

It's building up all the time. The young people have stopped going away. (Respondent no. 081)

'I got no sentiment for part-time fishermen.'
(Respondent no. 089)

Like the fishermen of Charleston, those of Grates Cove are bemused by a licensing system that distinguishes between full- and part-time fishermen. As all inshore fishing is seasonal, the distinction between full-time and part-time seems to them to make little sense:

I don't know what a part-time fisherman is. (Respondent no. 079)

Either you're a fisherman, or you're not. (Respondent no. 077)

None the less, many of our Grates Cove respondents resented the fact that part-timers, who generally have other occupations, have the right to take the fish on which the full-time fishermen depend. Consequently, the majority of those interviewed were in favour of some system for distinguishing between part-timers and those who depend entirely on the sea for their livelihood:

Yes, there is a need for a licensing system, if it is strictly enforced. But it is not. You give me the authority to go to any wharf and ask for fishing licences and then you'll see how many fishermen will be left. The fishery officers have got that power and they don't use it. (Respondent no. 078)

That's gone to Jesus, this licence system. But they're going to have to keep on with that till they get rid of the part-timers altogether in another twenty years. (Respondent no. 095)

As in King's Cove, many Grates Cove residents were particularly disdainful of moonlighters – those who also had full-time jobs outside the fishery:

It should be that people with other jobs should not be allowed to sell their fish. They are only killing a living for another man. (Respondent no. 093)

They shouldn't be allowed to put out nets and take up fishing space of a man depending on it for a licence. He shouldn't hinder the people who are depending on it. (Respondent no. 089)

[They] shouldn't have other fellas coming in and taking the bread off the table. (Respondent no. 096)

I got no sentiment for part-time fishermen. If a man is lucky enough to get a job, he shouldn't be taking away from the poor bugger who's depending on it. (Respondent no. 089)

They saw licensing as beneficial because they hoped that it would ultimately be used as a mechanism to reduce or eliminate moonlighters. And they saw the union's efforts to ban part-timers from selling fish to plants until all full-timers had sold their catch as a first step in that process.

They got the law now as good as you can get it. A part-time fisherman cannot sell until the full-time fisherman has sold his. If the government put down their hand firm enough they should take the licences from part-time moonlighters. (Respondent no. 083)

As in King's Cove, schoolteachers were singled out for condemnation:

There is a lot of talk that there is teachers and other people have full-time licences. Now it should be taken away from these people and given to other people who are fishing for a living. The same should apply for salmon licences. (Respondent no. 093)

A full-time fisherman is at it for a living. And you get teachers and some who have other jobs and don't do it for a living. (Respondent no. 094)

My opinion is that a full-time fisherman is one who depends on it for a living. There's fellas around here who are fishing and are schoolteachers too. (Respondent no. 090)

A schoolteacher fishing in the summertime is what you call an asshole. (Respondent no. 077)

Despite this strong condemnation of moonlighters in general and moonlighting schoolteachers in particular, some respondents emphasized that moonlighters had never been a problem in Grates Cove. It may well be that the kind of boat needed to prosecute the fishery safely under the severe weather and sea conditions typical of the Grates Cove area is beyond the means of most part-timers. This would suggest that the vehemence of the opposition to moonlighters has more to do with ideology than with the reality of the situation. The example of Grates Cove illustrates that opposition to part-time fishermen may be strong even when part-timers do not present a significant threat with respect either to the hunt for fish or to the hunt for income from the sale of fish.

As we saw in our analysis of other communities, fishermen also object to the licensing system's allocation of protected-species licences. Some of Grates Cove respondents reiterated the now-familiar complaint about the difficulty of getting species licences. Many of those in favour of the licensing system argued strongly that full-time fishermen like themselves should be the only ones allowed to have protected-species licences:

Yes, a full-time fisherman should be allowed to get clear of his fish first and have first crack at any limited licences available. A lot of fishermen here are trying to get commercial [i.e., protected-species] fishing licences and they can't get it. (Respondent no. 078)

No one can get a crab licence without a letter from God. (Respondent no. 078)

Also familiar by now is a related protest, against the fact that some protected-species licences remain in the hands of older fishermen and others who have become part-timers, and the conviction that those licences should be transferred to full-time fishermen:

They should save them [i.e., protected-species licences] for full-time fishermen. There are fellas here drawing their old age pension with salmon and lobster licences. (Respondent no. 090)

There are people with salmon licences that aren't even fishing, and there are people like we that would like to get them. (Respondent no. 080)

We are after trying for everything. You can't get salmon, herring, or lobster licences. They say there is too many licences issued. In my opinion, too many part-time fishermen got these licences. (Respondent no. 090)

A part-timer is liable to be a schoolteacher, doctors, and ministers. A full-time fisherman should get [the licence] because it is his livelihood. The part-timer is only at it for a few dollars and he is taking money away from the full-timer. (Respondent no. 079)

In Grates Cove, however, strong emphasis was placed on another theme that was apparently of lesser significance in the communities described earlier – namely, that the right to fish for all species (without being required to obtain additional species licences) should be inherent in the definition of full-time status:

A full-time fisherman who operates a boat should have the right to fish any species, to get any type of licence he wants, if he can handle the work. (Respondent no. 094)

I consider every fisherman fishing for his livelihood should be entitled to the same rights and privileges. (Respondent no. 082)

I feel I should be able to go out there and do what I got to do to make a living. But, I don't feel I got the right to encroach on someone else's living. (Respondent no. 077)

[A full-time fisherman] should be able to catch anything that swims. If the plant will pack it, you should be able to catch it. (Respondent no. 077)

In addition, and somewhat to our surprise, there were some complaints about the manner in which the Grates Cove fishermen handled the 'draw' for salmon berths. As we noted earlier, although full- and part-time fishermen may apply for protected-species licences to catch salmon or lobster, only a relatively few such licences will be granted in any given community. Because of the small number of people involved, in most communities no formalized mechanism for allocating salmon-net locations exists. The fishermen who are lucky enough to obtain salmon licences simply set their nets in locations they believe to be promising.

Perhaps because more salmon licences are allocated in Grates Cove than in the communities discussed previously, salmon fishermen there have developed a 'draw,' or lottery, system for allocating the most sought-after salmon-net locations. This draw is similar to that developed for the allocation of cod-trap berths, which we will be discussing later. The locations of the most desirable salmon-net berths are written on pieces of paper and placed in a bag, from which each eligible fisherman in turn draws the name of the location where he will be permitted to set his salmon net during the coming season. This lottery system is intended to eliminate the inequity of having the most productive and hence most sought-after berths occupied by the same fisherman every year. It also serves to avoid the open conflict that might arise if all eligible fishermen rushed to occupy the most desirable locations as soon as the salmon fishing season opened.

However, according to some Grates Cove salmon fishermen, the salmon-berth draw, instead of reducing inequity, actually tends to increase it. This is so because there is no prohibition in the licensing regulations against salmon-licence holders fishing together and, in Grates Cove, that is precisely what happens. This means that one crew may have two or more licence holders and, consequently, two or more chances to obtain the most desired salmon-net berths. Such crews have a decided advantage over crews in which there is only one salmon-licence holder. In the words of one disgruntled fisherman who found himself relegated to a second-rate berth;

If all four men in a boat have salmon licences, all four could reach in the bag and draw a number-one berth. There should be only one [person] in the boat [who] can draw. (Respondent no. 016)

This situation arises because of the very different conceptions held by government officials and local fishermen concerning the nature of salmon-fishing licences. As with all fishery licences on the East Coast, government officials see salmon licences as the personal property of the fisherman – they licence the man and not the boat. However, in most cases, the fishermen who are granted such licences are already members of boat crews with whom they share their other fishery catch and the income from it. Seen from the perspective of the local fishermen, it would be churlish indeed for a member of a fishing crew to declare that, although he was willing to share his income from other species with his crew members, he would not share with them the considerable cash benefits that he could obtain from his salmon licence. Not only could such a strategy lead to a rift between him and those with whom he must cooperate in other types of fishing, it would also make little sense in terms of maximizing the 'insurance' value that, as we have already seen, is an important part of the fishery process. Few, if any, local fishermen comprehend the basis on which government officials allocate protected-species licences; those who have a licence one year live in some fear that they might lose it in subsequent years. Under such circumstances, the wisest course of action is to share the benefits of protected-species licences with one's boat mates.

The treatment of the protected-species licence as a collective asset works well in communities such as King's Cove, where few licences are granted. In such places it is unlikely that more than one member of the same crew will hold the same protected-species licence, and the sharing of income derived from the licence is therefore likely to result in greater income equality. In Grates Cove, however, where up to four members of a single crew have been known to hold salmon licences, this process is likely to create substantial inequalities of income, thereby defeating the very purpose for which the community developed the 'draw' system. There is understandable resentment, therefore, on the part of crews that have only one salmon licence, not to speak of those that have no salmon licence at all. From the basis of their community value system, it is difficult for these fishermen to comprehend a licensing system that would fly so directly in the face of their fundamental notions of equality and justice.

It might be argued that the apparently undesirable consequence of the salmon-berth draw is proof of the local community's inability to regulate access to its own commons. However, that would be too facile an interpretation. The source of the problem lies, in fact, in the divergence between the government's and local fishermen's perceptions of the kind of property that the licence, in and of itself, represents. Whereas government officials may believe that the licence, being issued to the man and not the boat, is a form of private property, in practice local fishermen share the benefits of that licence among the boat crew and thereby transform it into a form of collective property. Because most crews are made up of kin, the protected-species licence is commonly viewed as a family resource. As evidence of this, several of the younger fishermen we interviewed in the various communities indicated that they had caught a few salmon during the previous season on their father's licence. They seemed completely unaware that they were effectively admitting to an illegal activity, a violation of the state's licensing regulations. Of course, their actions were entirely consistent with the community's understanding of the licence as either family or crew property. In this context, the prevailing belief among senior fishermen that they should be allowed to pass their licences on to their sons when they retire is also completely understandable.[3]

'You make your traps to fit your berth.' (Respondent no. 079)

Thus far in our analysis, Grates Cove appears to be little more than a prosperous and more professional version of the two communities analysed in the last chapter. However, its distinctiveness as a community and a fishery lies in the way it has established a basis for community control of the fishery. This control is inextricably linked to the inherent conflict between trapboat longliner and crews, and between those who rely primarily on cod traps and those who depend mostly on gill nets for their livelihood.

As we have noted before, current state fishery policy is predicated on the belief that regulatory systems are either absent or ineffective in most, if not all, inshore fishing communities. The state's response to what it assumes to be the inevitable consequence in a common-property situation – unbridled competition that can only result in overfishing – is to regulate *access* to the fishery through a licensing system. However, it has not developed a regulatory system to effectively police the *process* of fishing *per se*. In contrast, Grates Cove has its own system of communal 'laws' for the regulation of both access to the fishery and the process of fishing itself.

Whereas no one in either Charleston or King's Cove recognized the presence of a community system regulating the location of fishing gear, everyone we interviewed in Grates Cove was aware of their community's regulatory system. This was largely because of the explicit and public nature of the annual draw in which cod-trap berths were allocated.

Usually around the fifteenth of January fishermen from here gather to draw for traps. (Respondent no. 088)

We draws for berths here every January for salmon nets and cod traps. (Respondent no. 083)

We got a trap-berth draw. We've been drawing for berths as long as I can remember. (Respondent no. 083)

The current system was inaugurated some thirty-five to forty years ago, so it is no wonder that many residents cannot remember a time when it did not exist. Before that, fishing crews competed with one another on a first-come–first-serve basis. The extreme weather conditions in the Grates Cove area in the early spring made the setting of salmon nets and cod traps at that time of year a dangerous undertaking. Moreover, Arctic ice-fields and icebergs are prevalent in this area during the spring and early summer. Consequently, many fishermen who had risked their lives to secure a good berth saw their nets, and their livelihood, destroyed by ice. One old skipper explained the introduction of the draw for berths as follows:

We've had trap draws for almost thirty-five years. See, a place up here called Heart Cove had deep-water berths and in early May you could take fish. The same crowd would be out there marking their place in winter and early spring. Some fellas put out mock traps in January. After a few rows over that, we tried a trap-berth draw. (Respondent no. 087)

A younger inshore fisherman gave this version of the process that existed long before he had begun to fish:

Before, people would be going out in the winter to set gear. So they decided to draw berths and now they wait till it's time to set out the traps. (Respondent no. 081)

The berth-draw system is regulated by the local Fisheries Committee, which is elected by all those who are eligible to participate in the draw. Grates Cove fishermen took pains to distinguish between this long-standing community committee and the more recently formed Union Committee, which is also elected but is none the less seen as lacking the power and the status of the former body:

There are two committees. The Fisheries Committee or Cod-trap Committee and the Union Committee. The Cod-trap Committee sets berth draws and if one fella's trap is too close to another fella's you can complain to the Cod-trap Committee and they can straighten it up. There are five on the committee. (Respondent no. 078)

In the thirty-five years since the inauguration of the berth draw, Grates Cove fishermen have developed a wide range of additional community regulations. Among the most important is a system of categorizing berths. In Grates Cove there has always been a greater number of berths than there are trap crews; even taking into account that most crews have two or three traps, there are still more than enough berths for all:

There are only about fifteen trap guys [i.e., crews] and there's got to be at least forty berths. (Respondent no. 077)

However, not all trap berths are the same. Some are 'number-one berths,' which generally yield a high return, while those of lesser quality are called 'number-two berths.' Most of the number-one berths are in deep water to the north of the community, whereas the others are in shallow water to the southeast. To accommodate this difference, Grates Cove has developed two rounds of trap-berth draws. All skippers can enter the draw for number-one berths, and once that draw is over, anyone with extra traps can enter the draw for any remaining prime berths and for those that are considered second-class.

Usually around the fifteenth of January, fishermen from here gather to draw for traps. They draw for number-one and number-two berths. The number-one berths are mostly located around Heart Cove – deep-water berths. (Respondent no. 088)

Probably half of our berths are deep-water berths – over twenty fathoms deep – and the remaining half is probably ten to twelve fathoms. Those deep berths are in Heart's Cove, below Breakheart Point – to the north of here. The shallow berths are to the south. (Respondent no. 082)

The reason that the draw is held so far in advance of the start of the fishing season is now apparent: The time between the January draw and the start of the trap fishery in June allows fishermen to shape their traps to fit the berths they draw:

They sets small traps and big traps here in Grates Cove – about half and half. You can only set small ones at one point here in Grates Cove – to the east – and big ones towards the north. When you draws for berths you [may] draw a small berth or a big berth. You make your traps to fit your berth. (Respondent no. 079)

If a fellow got a deep berth he got to make [his trap] deep, and if it is in shallow, he got to make it shallow. You take a piece out of your trap if you got a shallow berth and a big trap. You can switch berths with a fellow with a deep berth if you get a shallow berth and vice versa. (Respondent no. 082)

Still, even all that planning does not guarantee that a prime berth will be better than a number-two berth:

We got one deep-water berth and one shoal-water berth. We had 125 thousand pounds of cod in traps, and there was 62 thousand pounds in one and 63 thousand pounds in the other one. It was the same last year. (Respondent no. 096)

The community has also developed its own policy defining eligibility to draw for berths. To qualify, a person must have been a resident of Grates Cove for at least one full year:

After a person has lived in Grates Cove for one year they are allowed to enter the cod-trap draw, providing they have a cod trap and full-time licence. (Respondent no. 082)

If you were living here, the second year you be allowed to draw. (Respondent no. 085)

Those who have lived in the community less than a year and even those who do not live there at all are still allowed to set traps in the Grates Cove fishing grounds. However, they may place their traps only in locations that remain unoccupied after the trap season has begun:

Whatever is left over after the berth draw can be taken by newcomers. (Respondent no. 088)

You have to become a resident of the community and go over to the hall and put in your name. There is a fella I know who come down here and puts out a trap, but he can't draw. He gets what is left. (Respondent no. 090)

Significantly, the right to draw for traps is not inheritable. Thus, the son of one elderly fisherman who was on the verge of retiring tried to take his father's place in the draw, even though he now lived in nearby Old Perlican. As his father explained, he was refused permission to participate:

My son came down from Old Perlican to try to draw in my place, but they considered at the meeting that since he wasn't a citizen of Grates Cove, he couldn't draw. (Respondent no. 079)

Of even greater interest is the father's evaluation of the committee's decision:

This is a fair basis. Years ago you got out where the good berths were, putting marks [i.e., markers] down. Drawing for them is the right way. (Respondent no. 079)

The preceding discussion has demonstrated that Grates Cove has a highly formalized community-based system regulating access to the fishing grounds. But, as noted earlier, the most significant aspect of the Grates Cove community regulation system is that it also features clearly specified rules governing the process of fishing. The basic intent of such rules is to reduce, if not eliminate, the gear conflict that arises among those who use different means and technologies of production.

One element of the community's regulation of the fishing process is the declaration of a predetermined date on which salmon nets must be taken from the water. This is important because good salmon-net berths are also quite often good cod-trap berths. Because the cod-trap season is so short, if someone were to leave his salmon net in a cod-trap berth after the cod had started to 'run,' he would seriously limit the annual income of the holder of that cod-trap berth.

There's a draw for salmon berths. Approximately twenty berths. Some are trap berths and it's agreed that salmon nets come out when trapping starts. (Respondent no. 086)

However, of greatest significance is the regulation that all gill nets must be out of the inshore water by August 15 of each year. Mid-August is approximately the end of the cod-trap season in Grates Cove. After that period, most inshore fishermen in the community turn to handlining:

They have a law for the community that you take up gill nets the fifteenth of August, so everyone can go handlining. (Respondent no. 093)

We made a law here that gill nets be off the ground for handlining at the fifteenth of August. That is, on the inshore ground. (Respondent no. 085)

All nets within a mile offshore have to be off the [fishing] grounds by August fifteenth. It's a gentlemen's agreement here in Grates Cove. (Respondent no. 078)

We have an agreement that all gill nets be removed from the grounds by the fifteenth of August. That is usually around the time the fish will take to the bait. (Respondent no. 082)

Since gill nets are the primary technology used by longliners, the regulation just described constitutes a direct limitation on the right of longliners to fish inshore waters after the end of cod-trap season. By restricting the use of gill-nets to offshore waters, the regulation is designed to ensure that sufficient fish come near shore to be caught by handlines:

Gill nets out after fifteenth of August. Then handlining. This gives everyone a chance to make money in the fall of the year. It's been that way since the old peoples' day and everybody abides by it. (Respondent no. 077)

As the preceding quotation indicates, such regulations go back to 'the old peoples' day.' Prior to the introduction of the gill net, fishermen used a similar technology, known as the 'floating cod trap,' and Grates Cove fishermen had similar prohibitions against it:

Way back when I was a boy there was no gill nets, but [there were] floating cod traps. And there was a law that you weren't allowed to have a cod float out before the fifteenth of June and you had to have it in before the fifteenth of August. It's the same now for gill nets, and everybody will use the handlines. It stops fellas from blocking things off with gill nets, cause the cod-trap season is only four, five weeks long. (Respondent no. 082)

These community 'laws' demonstrate not only that community-based fisheries regulation is possible, but that at least some fishing communities are willing to protect those who use traditional forms of fishing technology from any unfair advantage that might accrue to those who have the benefit of newer technologies.

The state has not displayed a similar willingness to protect traditional methods, primarily, it would seem, because it has adopted the view that the newer technologies, associated with larger boats, are the more desirable option. Some Grates Cove residents are highly critical of the state for failing to adequately regulate the capelin fishery. They consider current quotas for large capelin seiners to be too high and, because cod feed on capelin, to be a threat to the cod-trap fishery on which they base their livelihood:

In five or six years you can see the difference cause the capelin aren't coming in. The seiners are taking up too many capelin ... If they don't catch all of the capelin, the fishing here will have a future. (Respondent no. 093)

You take fifteen or so seiners which go out of Old Perlican and they clean up the quota. They took the Conception Bay quota in two days. (Respondent no. 078)

The fishery has no future here if they keeps catching the capelin the way they've been. 'Cause the capelin is cleaned up and the [cod]fish got nothing to come in for. (Respondent no. 083)

The seiners are overfishing in the capelin. That is the reason the [cod]fish don't come in. The capelin is overfished by seiners from Old Perlican, Catalina, and up the bay everywhere. It is a city out there in the nights with the capelin seiners. (Respondent no. 090)

So strong are such sentiments that the Fisheries Committee of Grates Cove compiled a petition among the fishermen demanding that the capelin seine quotas be lowered:

We had a meeting here in Grates Cove in order to set up a petition to stop them from catching capelin, and they cut down a bit the year [this year]. [The head of the Fisheries Committee] went around with the petition and it was cut down [i.e., the quota was lowered]. The petition was presented to federal fisheries. All the fishermen I knows of signed the petition. (Respondent no. 093)

The community action in this case would appear to demonstrate not that inshore fishermen are opposed to state involvement in the regulation of fishery practice, but that they see the role of the state somewhat differently from the way the state appears to see it. Grates Cove fishermen want the government to make a more concerted effort to protect them from those who reside outside the community and who are beyond its control. From their perspective, the appropriate role of the state is not to restrict the local community-based fishery (since the community has long demonstrated its own ability to do this), but to control the broader framework of conditions within which their local fishery is conducted.

'Competition is the name of the game.' (Respondent no. 080)

We argued earlier in the book that fishermen compete for two scarce commodities: fish, and the income to be earned from the sale of fish. We emphasized that the two arenas of competition are, in most instances, analytically separable. Thus, with regard to state policy, licensing regulations govern the competition for fish, whereas unemployment-insurance regulations influence the competition for income from the sale of fish. In Grates Cove, there is also a distinctive community-based form of social regulation within each of these two arenas of competition. At the community level, the competition for fish is governed by the regulations controlling the use of gear that we have just described. The competition for income is governed by practices determining the price of fish and the nature of relationships with outside buyers. Here again, the different technologies employed in catching fish are a determining factor.

The competition between trapboat and gill-net fishermen extends beyond the fishing grounds; it is also present in the area of pricing, specifically with respect to the prices offered by fish plants for the fishermen's catches. At the time of our study, the plants were paying a higher price per pound for gill-net fish, than for trap fish, despite most trapboat fishermen's conviction that their fish was fresher and, therefore, the superior product. As noted earlier, trap fish remain alive until the trap is hauled, whereas gill-net fish drown and may remain in the nets for some days before they are harvested. The differential pricing program infuriated many of the trapboat fishermen:

There is one thing I see as a crime. I go out there and hoist my trap and I bring my fish back in alive, and the man next to me brings in fish that have been in the gill nets three or four days that are rotten, and they get more. That's something I'd like to see the union fight. (Respondents no. 081)

They talks about high-class fish around here, and some fella with a gill net gets more than what we are selling it. It should be the other way round. I can't see the reason why they got it that way. (Respondent no. 092)

Now the gill net-fish could be out there for weeks and you get more for that than for a live fish from the trap. You should get more money for the live fish from the trap 'cause he is fresh. (Respondent no. 079)

This is an often-heard complaint among rural Newfoundland fishermen, because union contracts with fish buyers usually stipulate a higher price for gill net-fish and give the best price of all to fish caught by handline.

The longliner fishermen in our sample were quick to counter charges that their fish was inferior. They contended that trap-caught fish were not only smaller, but also a softer fish, because they had been glutted with capelin and had been taken in warmer shoal water. They argued that this caused them to spoil more quickly and also made them harder to fillet. They maintained that, as a result, trap fish yielded fewer pounds of fillets per pound of 'round' fish than did gill-net fish:

If anyone told me a few years ago when I was trapping that a trap fish wasn't better than a gill-net fish, I'd have told them they was a liar. But I worked in the fish plant for two years and I saw that gill-net fish keep better. Trapped fish are soft. They get soft when the glut is on [i.e., when they are full of capelin]. (Respondent no. 094)

The lower price is for trap fish. From what I heard is that a trap fish is a small fish, a shoal-water fish who will spoil faster. A trap fish will never give you cold hands when you are gutting them. (Respondent no. 085)

One longliner operator suggested openly that complaints about the quality of gill-net fish had less to do with the fish than with the competition between fishermen using different types of technology. He noted that longliner operators complain just as vociferously about draggers taking fish from their offshore fishing grounds:

In order to avoid conflict, you'd have to go into two classes of inshore fishing – shoal inshore and deep inshore. It's the same way with us [i.e., gill-net] fellas. We complain about the draggers. (Respondent no. 094)

However, at least one trap fisherman was not convinced by such claims:

The plant crowd says the trap fish gets soft quicker than the gill-net fish and there is not much that you can do with it. But I wouldn't eat a gill-net fish. The majority of people at the wharf wouldn't get a gill-net fish to eat. They get a trap fish. (Respondent no. 096)

Whatever the legitimacy of these rival claims, some trap fishermen argued that the buyers should introduce a new system of grading fish based on the freshness, cleanliness, and overall quality of the actual fish as it was delivered for sale, rather than on the way in which it was caught:

There should be an A, B, C, quality of fish, no matter where he [i.e., the fish] comes from. (Respondent no. 080)

No such system was in use at the time of our study.

However, the fishermen of Grates Cove have developed at least one 'community-oriented' way of exercising some control over the competition between fishermen in the sale of fish. It involves turning the competition back on the fish buyers themselves. Representatives of two rival fish plants located in other communities compete to buy fish in Grates Cove. Grates Cove fishermen make sure that they divide their sales between the two buyers:

We sell one week to [one firm] and the next week to [the other firm] to keep competition alive. I say the majority in Grates Cove does that. (Respondent no. 083)

They argue that, by doing so, they have managed to raise the price they receive for their fish:

The two boys are there so we sells to one one week and to the other the next week to keep 'em here. One time there was only [one fish buyer] here. The price has gone up since the two buyers have been here.

Even more important, however, the fishermen contend that such a strategy ensures that there is always a buyer in the community and that each buyer acts in the interest of the fishermen:

We alternate [buyers] from week to week to keep two buyers in this area. Competition is the name of the game. (Respondent no. 080)

We sell one week to [one buyer] and the next week to [the other buyer]. We try to get the competition. Otherwise one buyer would do what he wanted to us. (Respondent no. 081)

Put more abstractly, if competition for income is a basic aspect of fishing – the proverbial 'name of the game' – then Grates Cove fishermen are doing what they can to structure that competition in their own favour. The outcome of their strategy is one more example of community 'norms and practices' having altered the nature of the competition and the potential for conflict that is inherent in the fishery. Fishing may indeed remain a 'game' and a 'gamble':

The money's not in fish like most other jobs, but you get by on it. It's a gamble, fishing is. (Respondent no. 080)

Nevertheless, the efforts of the Grates Cove fishermen to affect the activities of the buyers is an important example of a community that is actively attempting to take control of its own fishery.

'Every inch, every inch of Grates Cove, I love it.'
(Respondent no. 089)

Just as Grates Cove fishermen like their work, so too do they like their community. Some, like the respondent quoted in the heading above and the one quoted below, were ecstatic in their evaluations:

There is no other place in the world that I would like any better. (Respondent no. 079)

All those we interviewed indicated that they liked living in the community, and twenty-one of twenty-two respondents believed that it was a good place to raise children and that their neighbours cared about the place and were generally hard-working.[4] As two respondents independently noted,

You got to be hard-working to live here. (Respondent nos. 092 and 096)

Similarly, as we can see in Table 6.1, the great majority had praise for the teachers and schools in the community and for the medical care available in it. Few felt that living in Grates Cove meant that they had to do without a lot.

Despite this, a strong level of ambivalence became apparent when our interviewees were asked about the future of their community and its way of life. Although the vast majority felt that the community had a good future, they were far less certain about the future of the fishery and felt there was no other work available. Consequently, they were divided on whether the inshore fishery was good for young people and whether their own children should settle in Grates Cove. Some were positively disposed to the 'gamble':

TABLE 6.1
Grates Cove respondents' views of their community and its future

	Yes	No	Uncertain
Evaluation of community			
Like living here	22	0	0
Good place to raise children	21	1	0
People here generally hard-working	21	1	0
Leaders are good, capable people	7	0	0
People here care about the place	21	0	1
Evaluation of community services			
Good schools and teachers	18	0	3
Good medical care	19	1	2
Enough to do in spare time	11	0	0
People here do without a lot	5	17	0
Evaluation of employment opportunities			
Good place to find work	2	20	0
Fishery has a future here	12	3	7
Fish plant here has a future	21	0	1
Inshore fishery good for young people	8	12	2
Evaluation of community's future			
Children should settle here	6	7	3
Community has a good future	19	0	3

Note: There were 22 respondents. Where responses for any given item do not total 22, the balance represents abstentions.

It's a gamble, with licensing and everything. But if he wants to take the gamble, yes sure. (Respondent no. 086)

It's as good now as it ever was, as long as the fishery keeps going. (Respondent no. 075)

The way things are going, I'd say it's just as good as anything else. (Respondent no. 076)

Others considered the gamble too great:

No, I wouldn't advise a young fella to go into it cause there are too many draggers on the go. As far as I can figure they [the government] don't want a inshore fishery. If the draggers had their way they would all be inside Trinity Bay ... If a young fella had education and a half-decent job, I wouldn't advise it. (Respondent no. 087)

If you had no [i.e., any] sense you wouldn't go at it. (Respondent no. 083)

It should be remembered that many of the respondents were themselves quite young, and had small children. Even though many of them had left other places and other occupations to return to Grates Cove and felt that they were doing reasonably well, they were unsure that the fishery would be able to sustain the next generation.

FERMEUSE

Fermeuse lies about 120 kilometres due south of St John's, on a coast known locally as the Southern Shore.[5] Despite this relatively short distance, the region is in many ways isolated from St John's and has its own distinctive culture and a unique social character. The entire coast has many long inlets that afford excellent harbours. Fermeuse is situated at the end of one such inlet, which it shares with the smaller community of Port Kirwin. Port Kirwin is located on a small indentation closer to the mouth of the inlet. Fishery licensing records combine Fermeuse and Port Kirwin. Consequently, our Fermeuse sample includes some respondents from Port Kirwin as well. Although there is some rivalry between Fermeuse and Port Kirwin fishermen, it pales in comparison with the intense antagonism that exists between Fermeuse and Renews, the neighbouring community that occupies a similar inlet some ten kilometres south. This intercommunity rivalry provides a leitmotif for our discussion in this section.

The harbours at Fermeuse and Renews were among the first to be occupied by seasonal fishermen from Europe. Fermeuse appears on a 1516 Portuguese sailing map as 'Formoso,' which means 'beautiful.' Renews is mentioned by Jacques Cartier, in his account of his 1536 voyages, as 'Rougnouse', which may derive from the French *rogneux* meaning 'scabby' or 'covered with kelp and shells,' or from the Portuguese word for 'sandy' (Seary 1967, 259). Both locations were noteworthy not only for their harbours but because the surrounding land slopes to the sea and there was always a plentiful supply of fresh water and firewood.

The same characteristics later attracted colonists. In 1622, Sir George

Calvert founded one of the first colonies in the New World at Ferryland, the next harbour to the north.[6] Lord Falkland's colonial territory of South Falkland was awarded in 1623 and included both Fermeuse and Renews. In a published pamphlet he advertised for sale '[fishing] stage room for 2 Rooms at Renews and Fermeuse' (Prowse 1895, 120), but the response to the proposal was poor and the colony was not occupied. Beginning in 1633, the English Star Chamber formally turned the governance of Newfoundland over to 'fishing admirals.' Each harbour had its own admiral, who was the captain of the first fishing ship to enter the port that year (Thoms 1967a, 530). The diaries of James Young, a doctor and adventurer who visited Newfoundland with the fishing fleet in 1663 and 1664, refer to an 'admiral's place' in Renews Harbour and a 'vice-admiral's place' in Fermeuse Harbour (Seary 1967, 261).

The area continues to be noted in various documents from subsequent years. A 1696 account of the French attack on St John's and other nearby communities indicates that the French 'now intended to inhabit Renews, having already fortified that place' (Prowse 1895, 230). Both 'Fermouse' and 'Renowse' are listed as belonging to the Ferryland judicial district in 1732 (Prowse 1895, 301). In 1780, Renews is described as being fortified by six cannons, which the residents used to hold off a marauding privateer (Prowse 1895, 230). In 1817 a mob of Renews residents, led by the local blacksmith, is reported to have attacked two ships frozen in the harbour ice, apparently in a desperate attempt to obtain the provisions they needed to save themselves from starving (Prowse 1895, 405). The early residents of Fermeuse and Renews were poor fishermen, who are not individually identified in the historical record. The first record of families now living in Fermeuse dates from the early 1800s. Constantine O'Neill, born in 1797 in County Kilkenny in Ireland, died in Fermeuse in 1810. Andrew and Robert Oates, John and Thomas Fennelly, John Shaughnessy, and John and Matthew Brophy are recorded in the 1871 census as 'fishermen of Fermeuse' (Seary 1977).

Census data for 1986 show 546 persons residing in Fermeuse, which makes it almost twice the size of Grates Cove. Another 142 people lived in Port Kirwin. Our list of licence holders showed that, in the two communities combined, there were 126 men (and no women) with general fishing licences, of whom 68 were full-time and 58 were part-time fishermen. In Grates Cove an astonishing 1 out of every 3.8 residents held some kind of fishing licence; in Fermeuse, 1 out of every 5.5 did. The big difference between the two communities was in the ratio of full-time licence holders to total population. In Grates Cove, 1 in every 4.8 residents held a full-time licence, while in Fermeuse the ratio was only 1 in 10.1. Indeed, there were almost as many full-time licence holders in Grates Cove as in Fermeuse, despite the considerable difference in the size of the two communities. There were also eleven salmon-licence holders and three lobster-licence holders in Fermeuse, including two fishermen who held one of each. All such species-licence holders were full-time fishermen. Although Fermeuse is not the last community discussed in this book, it is the last community in which we interviewed residents, and we were already short of both time and money. Consequently, we interviewed only twenty fishermen, but, to maximize the amount of evidence we received, we chose all of them randomly from the list of full-time licence holders. Fifteen of these were from the list of fishermen who held only general fishing licences, and five were from the list of those who held a species licence for lobster or salmon.

The reasons for choosing Fermeuse for inclusion in this study were outlined in chapter 1. Fermeuse had already been studied in the early 1970s by Kent Martin, who had focused in particular on the way local custom governed the delineation of 'fishery space' (Martin 1979). Because of the similarity between the concerns of Martin's study and those of our own, we are now able to offer something of a diachronic perspective on community control and competition as manifested in Fermeuse at two points in time nearly fifteen years apart.

Martin (1979, 285) argued that Fermeuse fishermen had long since developed two strategies for regulating use of the fishing grounds:

First, symbolic relations are established with neighbouring communities whose fishing grounds complement the deficiencies in Fermeuse grounds at varying points in the fishing cycle ... Thus, Fermeuse trap fishermen regularly set their second trap (most crews have two traps) in unused areas belonging to the neighbouring community of Aquaforte, and Aquaforte fishermen frequent the Fermeuse offshore banks during the fall fishery to compensate for the virtual nonexistence of such areas in their own community.

Second, Fermeuse fishermen have divided their own fishing grounds, as have many inshore fishing communities, by setting aside certain areas (usually the most productive) for the exclusive use of certain technologies.

Martin paints a vivid picture of a community in harmony with its ecological base, a community whose dominant goal is to 'balance the existing fishing space with the number of fishermen' (1979, 277). However, it is worth noting his observation that 'such 'sanctuaries' have not been established without dissension. Rather, they are usually the upshot of protracted and often very heated disputes over which technopolitical faction has the 'right' to utilize prime fishing areas. The faction that prevailed in these disputes was usually the one to which the bulk of the local fishermen were most heavily committed in terms of gear, crew organization, and exploitive expertise' (285).

Martin also describes a situation in which competition is carefully controlled so as to 'minimize social conflict' (1979, 277): 'A premium is placed on maintaining ostensibly cordial relations with competitors (who are also one's neighbours) ... The idea is to avoid overtly aggressive competition but nevertheless to manoeuvre a competitor into 'doing the job himself' ... The fisherman who expects his neighbour to help him find the major fish concentrations will end up with a poor season indeed' (290–1). He stresses that 'the intense competition that takes place on the fishing ground ... is not allowed public acknowledgement ashore' (290).

Our data suggest that the controlled and civilized competition Martin described is at times quite unstable. Indeed, we found a community divided within itself and at war with its neighbouring community. Intense competition between those who used different types of gear also militated against internal solidarity.

'There's not a spare berth left around here.' (Respondent no. 100)

The organization of fishing work in Fermeuse was the most complex that we had yet encountered. As in many other fishing communities, in the course of any one season an individual fisherman might employ a wide range of technologies, including traps, gill nets, and handlines. But, unlike the case in most of the other communities we had studied, the available fishing grounds around Fermeuse are exceptionally small and overlap those of other nearby communities. This is the cause of the intense competition in the area not only among fishermen who use different technologies but also among those who use the same technology. To complicate matters further, Fermeuse has developed a fishing-crew structure that is more complex than any we had previously observed. In most other communities, fishing crews tend to stay together throughout the season (and even from season to season), first using traps, then nets, and, finally, handlines as the season progresses and as the quantity and condition of the fish change. However, in Fermeuse the crew structure tends to change throughout the year. Thus, a man may fish with one crew during the trap season in early summer and switch to a different crew to use gill nets later in the season. Then, as fall approaches, he may take his own boat and engage in the handline fishery.

As in many inshore communities, it is the trap fishery that is the dominant, and, generally, the most profitable form of fishing. Primarily because of this, Fermeuse, like many other communities, has instituted a draw for berths. It seems ironic that communities have chosen to regulate the fishery – an enterprise that is very much a gamble in itself – by means of a lottery. Instead of providing a measure of stability to the fishery, such a system institutionalizes the workings of chance.

Some of our Fermeuse interviewees claimed that the community draw for berths had always been in place:

We have a cod-trap committee and we hold a meeting to decide the trap berths ... This draw has been going on for as long as I can remember. (Respondent no. 099, age 59)

Others spoke of a time before the introduction of the draw system when berths were regarded as family private property and were handed down from father to son.

In the old days, berths were passed on from father to son, so the only way you could get in was for someone to give it up or die. (Respondent no. 104)

If this was indeed the case, then the community draw transformed a private-property system into what is essentially a community-regulated common-property system. Since most research on common-property systems has tended to define them as preceding the creation of the institution of private property, this is a significant observation. It suggests that the transformation of property from one type to another does not necessarily follow the linear progression that has generally been assumed.

In the majority of fishing communities, there tends to be a relative abundance of trap berths, even if some of them are not as good as others. In such communities, the available trap berths are generally identified as 'number-one' and 'number-two' berths, and two draws are held to give each crew an opportunity to obtain both a first- and a second-grade berth. This system allows each crew to increase its income through the use of two traps, while improving the odds against having a disastrous trap season as a result of the failure of the fishery in one particular area.

Presumably, the draw system was instituted in Fermeuse for similar reasons – to average out opportunities and incomes, and to give everyone an even chance. But, as noted earlier, Fermeuse differs from other communities in that its fishing grounds are so small that they contain relatively few potential trap berths:

We have a cod-trap draw. There are only ten berths available and there are ten crews fishing 'em. (Respondent no. 097)

We have a cod-trap committee and we held a meeting to decide on the trap berths. We have only eleven berths in this area and we have eleven draws. (Respondent no. 099)

Moreover, some of these ten (or eleven) berths are generally agreed to be of lesser quality than others that are included in the same two draws:

That's one of the problems b'y. There's only eleven berths around here, and some of them are no good. And there are eleven boats, so that's only one draw. So if you get a bad berth, which we did this year, that's tough. (Respondent no. 100)

We didn't set the year, because we had a poor berth. You got to have your trap out the fifteenth of June or someone else can get it for that year. (Respondent no. 098)

But the problems of cod-trap fishing in Fermeuse are not limited to those who draw a bad trap berth: Even worse off are those who do not have the right to draw for a trap at all. As the community grew, the number of people fishing from Fermeuse came to exceed the number of available berths, and the right to participate in the draw came to be limited to those who had fished trap berths in the preceding year:

If a fella has a berth one year, he can enter the draw next year. That berth is licensed and protected by fisheries [i.e., the Department of Fisheries]. (Respondent no. 097)

We got twelve berths here, and so long as you set your trap that year, you can enter the draw the next year. (Respondent no. 103)

However, the right to occupy a berth and, hence, to participate in the draw is inheritable by other members of a fishing crew when their skipper/boat owner retires or gives up fishing:

When I retires, I'm going to get the berth changed over to my brother – also my boat and salmon licences. (Respondent no. 099)

One fella, he has a trap, but he went dragging on the west coast [of Newfoundland]. So he had a cousin of his [from another community] come in and operate his trap. (Respondent no. 100)

As might be expected, the right of trap-berth inheritance contributes to a process of crew migration wherein crew members willingly move from a crew that lacks the right to draw for a trap berth to one that has it:

I go with fellas who get a trap berth. (Respondent no. 108)

It [i.e., with whom one fishes] depends on the trap berths. People like to get with a fella with a good berth. (Respondent no. 103)

In order for fishermen who are not members of existing trap crews to become eligible for the trap-berth draw, one of the existing crews must give up the trap fishery entirely:

No, there are no berths available now, someone would have to drop out for someone else to get in. (Respondent no. 115)

Right now, someone got to back out of the berth before you can draw for one. (Respondent no. 103)

The practices just described indicate that, in Fermeuse, the nature of property in the fishery has changed once more. In an earlier time, the right to 'own' a trap berth was transformed from private property to a form of community common property with the introduction of the trap-berth draw. Since that time, however, the right to participate in the draw has become, *ipso facto*, a right to exclude others from participating in the trap fishery. And, as we noted earlier, a right to exclude or to limit access, more than the mere right of possession, is the defining mark of property ownership. In a sense then, the ten or eleven crews that participate in the trap-berth draw have turned that fishery into a form of 'joint' or 'corporate' property that is inaccessible to the majority of fishermen in the community. As such, it is an example of yet another modification, or another type, of property relationship.

Indeed, in addition to waiting for a trap berth to be forfeited, newcomers must meet a series of eligibility requirements before they will be considered for the draw:

In order to get into the draw you also need a boat twenty-two feet or more in length, have it registered, and be a full-time fisherman. (Respondent no. 097)

For a newcomer, he'd have to set a trap in a berth for a year in a location not in the primary berths. [This is] to show people he is serious. He could enter the draw the next year if someone drops out. (Respondent no. 101)

Given these circumstances, it is little wonder that a level of resentment built up in the community against those who securely held the right to draw for trap berths. This resentment boiled over in the early 1980s, resulting in a legal action by a fisherman without a trap berth to demand that he be given the right to enter the draw. To be sure, if the number of good berths was really as small as all parties seemed to believe, the wisdom of this particular demand seems questionable. After all, if granted, the fisherman's action would not significantly open up access to the trap-berth fishery in Fermeuse. The challenge was, in fact, successful, with the result that one more inadequate location was added to the pool of berths and one more crew gained the right to participate in the draw.[7] As one trap-berth holder who was opposed to the newcomer's position explained,

There was another from the place who wanted a berth, and we tried to stop it. But they put it in court and he got it. We [i.e., the respondent's crew] ended up with a bad berth, and didn't bother going at it [i.e., fishing] that year. (Respondent no. 109)

Of course, the fisherman who initiated the action had a different view of the situation and its outcome:

We had to get a lawyer and get the law changed. [The judge decided that] they either had to let us draw, or each crew [had to] give up a part of their catch.

We could have taken the crew that made the most money that year and called that our shares, and taken some from all the crews to make it up. They had to put in a second-draw berth. (Respondent no. 104)

Perhaps the most significant aspect of the situation was the willingness of the courts to make judicial decisions that affected traditional fishing rights. The court's intervention struck at the core of the community's right to regulate access to its own fishing grounds. Many in the community believed that, rather than establishing the legal legitimacy of community regulations, the court's decision served to undermine the power and authority of community-based 'laws.' It established that community property rights and access regulations were subject to external legal challenge. As a result, many Fermeuse residents lost confidence both in the community committee that had traditionally been responsible for regulating the fishing grounds and in the authority of the traditional regulations themselves:

It is no good making harbour laws, 'cause it don't stand up with the Fisheries [Department] in St John's. (Respondent no. 099)

Indeed, the court's decision further divided and disillusioned the bulk of the fishermen:

Because of the upheaval over the extra trap berth that was brought in a couple of years ago, our fishermen's committee has not been together ... From my point of view, we got disillusioned when we couldn't prevent federal Fisheries from putting that extra berth in. Now, if we can't have impact upon a minor decision like that, I don't know ... (Respondent no. 106)

'I didn't earn enough with the trapping crew, so I switched to a netting and trawling crew.' (Respondent no. 110)

It is clear that, in Fermeuse, other forms of the fishery are, or at least have been, secondary to the trap-berth fishery. For example, gill nets and trawls, according to both community custom and Fisheries Department regulation, must not be set within fifty feet (roughly fifteen metres) of a trap. Indeed, it is the importance of the trap fishery, coupled with the small number of trap berths available, that has given rise to Fermeuse's highly fluid crew structure. Whereas in other communities fishermen fish throughout the season, 'the father and sons and the two sons together,' that is less likely to happen in Fermeuse. There, each man negotiates on his own to try and join one of the trap-berth crews.

For example, one respondent fished one season with a trap crew from the nearby community of Aquaforte (none of the members of the crew was related to him), and the next season with a crew from Fermeuse. Another fisherman spent four years fishing with a trap crew from Fermeuse, one year with a crew from Aquaforte, and three years with a crew from Kingman's Cove. Only in the case of the Fermeuse crew was he related to any of the other crew members. Yet another fisherman had fished for several years with an Aquaforte crew. In this, our findings complement those of Kent Martin (1979). He suggested that the underutilized fishing grounds at Aquaforte provided locations for some Fermeuse fishermen to set their traps. Our findings suggest that, approximately a decade later, most of those locations had come to be occupied by Aquaforte crews, who were willing to sign Fermeuse residents on as additional skilled labour. In the process, however, the status of Fermeuse fishermen declined.

Such crew switching is not limited to realignments of trap crews from season to season. For most fishermen, it also occurs within a single season, as they move from one technology to another. For example, one fisherman began the fishing season catching salmon with two other men. He deemed himself lucky, as they had 'two salmon licences in the boat' (Respondent no. 100). At the end of the salmon season this crew was expanded to include two more men and they participated as a crew of five in the trap fishery. At the end of the trapping season, our respondent remained with the original two partners to form a handlining crew. Another respondent (no. 101) did not engage in the salmon fishery at all, as he had no licence and was unable to join up with anyone who did. He began trap fishing with four other men in early June. At the end of the trapping season, in early August, he went handlining with two of these men. A third respondent described his changing crew composition as follows:

The first part of the year we gill-nets, and there was two of us gill-netting. There is four of us with the trap in the summer. We [i.e., the respondent and his gill-netting partner] have another fellow with us trawling. (Respondent no. 105)

The changes described by these respondents may seem like only minor additions and subtractions to the same basic crew. But their implications are not insignificant from the perspective of the fishermen's income. It should be noted that very different financial arrangements cover the various stages of each fishing season. For example, the first respondent described above (Respondent no. 100) received a one-sixth share of the money from the 65,000 pounds (roughly 30,000 kilograms) of cod sold by his trap crew (that is, one-sixth for each of five men and a share for the boat). However, he received one-third of the cash from the sale of 85,000 pounds (38,500 kilograms) of cod caught handlining (that is, three equal shares for the three men in the boat). The potential difference between the productivities of two different trap berths is illustrated by a comparison of the data supplied by this respondent and those supplied by the second respondent described above (Respondent no. 101). That person also received a one-sixth share of the money earned from the trap-caught fish, but his crew's catch totalled more than 400,000 pounds (roughly 180,000 kilograms). His handlining crew, however, caught only 15,000 pounds (6,800 kilograms), from which he received a third of the income. The income of the third respondent described above (Respondent no. 105) is even more complex. He received a 50 per cent share of the 100,000 pounds (45,300 kilograms) of fish that he and his partner caught by gill-netting, but his trap fishery was very poor, with a catch of only 20,000 pounds (900 kilograms), to be divided four ways. However, he and his two companions caught 150,000 pounds (about 68,000 kilograms) of cod trawling and another 50,000 pounds (22,500 kilograms) handlining, and each received a third of the income from the sale of the catches from each of these two types of technologies. When one takes into account that the price paid for cod varies depending on the technology used to catch it, the 'calculus' involved in deriving a living from fishing starts to take on an astounding complexity.

'Any person, priest or mountie, can get a licence
to fish part time or full time.'
(Respondent no. 100)

Fermeuse displays virtually all of the internal conflicts that we have observed in the communities previously described. They include the seemingly inevitable differences between full-time and part-time fishermen and considerable conflict among those using different technologies of production for catching fish. As we noted in our discussion of Grates Cove, it is these two forms of the division of labour (rather than the divisions of social class that have generally received more attention from social scientists) that are of greatest relevance to the fishery and that have the potential to divide all fishermen. In this section we examine how these two divisions of labour sow the seeds of the intracommunity conflicts that divide the fishermen of Fermeuse. In the following section we will examine how these same two divisions of labour also create the basis for the strong intercommunity conflicts that exist between Fermeuse fishermen and those of neighbouring Renews.

In Fermeuse we interviewed only full-time fishermen, and it is not particularly surprising that virtually every respondent was critical of part-timers. This attitude is reflected in the following comments:

I think part-time fishermen should be cut out altogether. (Respondent no. 108)

To be drastic, you'd tell the guys that's got the (non-fishing) job to get out of fishing, and let the full-time fishermen do it. (Respondent no. 101)

For some, such condemnation was the product of their sincere belief that there were simply too many fishermen attempting to derive a living from the limited fish stocks found in the Fermeuse area:

There is too much fishermen and not enough fish. (Respondent no. 115)

Right now, there are too many boats chasing too few fish. (Respondent no. 098)

Many other complained bitterly that the part-time fishermen also had full-time jobs from which they were able to make a good living, whereas full-time fishermen had to compete with them for the limited income there was to be wrested from the sea:

There are a lot of fellas here – we calls 'em moonlighters – who are getting in your way. They don't depend on it for a living. (Respondent no. 109)

A guy with another job shouldn't be able to fish. There's guys come down on their vacations, and if you get forty or fifty of these guys it cuts in [to full-time incomes]. And they don't care ... I think a part-time fisherman comes out of it better in the long run. The union does the same for him as for me, and he's got a job to go back to. (Respondent no. 104)

There's firemen and everything fishing. A man who's got a full-time job shouldn't have the same rights as a fisherman who's depending on it. (Respondent no. 102)

The only difference is that a part-time fella got another job. It's only a sport for him ... If there's any money available it should be there to help out the full-time fishermen. (Respondent no. 105)

As a result, many saw no significant difference between part-time and full-time status:

Well, right now it makes no difference, 'cause I don't see no advantage to [full-time status]. You can go out tomorrow and catch a pile of fish just like I can. (Respondent no. 097)

I think the system is a farce, 'cause they say they designate some people as part-time and the full-time shouldn't have to contend with them when making their livelihood. [Yet] they could be in prime areas where I could be fishing. (Respondent no. 106)

The only benefit to the licensing is they [i.e., the government] know who's part-time and who's full-time. But they're not doing a damn thing about restricting the moonlighters. (Respondent no. 100)

A couple of respondents did mention the primary difference between part-time and full-time fishermen, which is the right to sell fish during a glut:

I don't see any difference right now, except a full-time fisherman can sell his fish first in a glut. (Respondent no. 097)

A full-time fisherman draws UI in the winter. (Respondent no. 117)

Some suggested restrictions that they felt should be put on part-time fishermen:

A part-time fella should not be allowed to trap or gill-net. He should be only allowed to handline and jig. If the market's poor, the part-time fella should not be allowed to sell his fish. (Respondent no. 105)

Part-time fishermen should not be allowed a cod-trap licence and should not be allowed to set their nets in areas where full-time fishermen have set them. (Respondent no. 113)

Other complained of lax enforcement of the regulations currently in effect:

I know of people who haven't used a licence for two years, and they're still selling fish, and people are still buying it. And I wonder how many people are doing it. I wonder if the government can know? (Respondent no. 112)

There are fishermen around here who fish without fishing licences – them that are only just at it. (Respondent no. 114)

This conflict between full- and part-time fishermen was interwoven with the issues surrounding whether part-time fishermen should be able to hold protected-species licences. Although the majority had nothing to say about the matter, some spoke vehemently against the practice:

Salmon are scarce and they should give licences only to full-time fishermen. (Respondent no. 117)

There is a scarcity of salmon and lobster. The licences should go to bona fide fishermen. (Respondent no. 109)

However, others felt that the licences were of such little value that they were hardly worth having:

The salmon fishery around here don't mean much anyway. You might make a dollar and you may not. (Respondent no. 100)

I've no reason to apply for one. I'm busy enough. (Respondent no. 110)

The other major internal conflict concerned the type of technology used to catch fish. The primary concern was about the use of gill nets. Fermeuse fishermen have developed their own regulations governing where such nets can be placed. As in other communities, however, a considerable number of residents still believed that gill nets should be banned from the fishery. They were seen as destroying the market, primarily because the quality of fish caught in them was below that demanded by the consumer:

The gill nets is the worst thing brought into the fishery. The fish from gill nets is killing the market. They should be banned. This summer they got gill nets all around the cod traps, that is blocking fish from the trap. Last fall, there was more money [per pound] they got from gill net fish than from better-quality fish from handlines. (Respondent no. 103)

The gill net is killing the market I think. Trawls is the right way to do it. Certainly no one here would ever eat a fish from a net. But you got to do it [i.e., use gill nets] because everyone else is doing it. If they cut it out for everybody, everybody would have an equal chance to catch fish – fresh fish. It would be much better for the market. (Respondent no. 112)

The gill nets, my friend, is the ruination of the fishery. If the people who bought fish could see gill-net fish, they wouldn't buy it. (Respondent no. 103)

There was also consternation that the gill nets were 'blocking' access to the cod traps, thereby 'destroying' that mode of fishing:

I'm all for the trawls, but you take a guy with thirty-five nets, he could ruin the fishery for lots of fishermen. There are lots of gill nets in Albert's Cove, or Port Kirwin as they call it now. Last year we took in thousands of fish sometimes. Now this year not a shagging thing because of the gill nets. (Respondent no. 103)

Even some of those who used gill nets spoke against them:

There's one thing I don't understand. A trapped fish is the best fish there is. It's a hard thing to look at when you get twenty cents for a large trapped fish and a guy comes with fish [that have been] three or four days in a gill net and gets twenty-four cents. (Respondent no. 104)

At the same time, people seemed resigned to the fact that little could be done about the existing situation:

They were talking about getting a petition here in order to keep the gill nets off two miles [from shore]. But nothing came of it. There is not much you can do about the gill nets. (Respondent no. 099)

'The only big issue here is the Renews' Rock area.'
(Respondent no. 097)

Indeed, at the time that we were interviewing in Fermeuse, the 'big issue' centred on the right to fish at Renews' Rock. 'Renews' Rock' is the local name for a small underwater shoal, about five to six fathoms deep, surrounding a 'rock' that lies offshore from Renews but within reach of the small boats of Fermeuse. Almost every interview we conducted in the community eventually settled either on the right of Fermeuse fishermen to fish in the Renews' Rock area or on the type of technology that was legally allowed to be used there, or both. Some respondents either implied or stated outright that the controversy was rooted in some fundamental difference between the communities of Fermeuse and Renews.

There's two towns that just couldn't be more different. (Respondent no. 112)

Nowhere else did we encounter a controversy that pitted one community against another in a major battle over fishing property rights and the fishing practices resulting from them. The Renews' Rock conflict brings into the open the many divisions and conflicts that usually remain hidden within and between fishing communities. It calls into question the endurance of the community conflict control that was the focus of Martin's analysis (1979). Although the 'symbiotic control between communities' that he described may have existed in the past, it most certainly does not exist now. Moreover, the premium on 'minimizing social conflict' and on maintaining 'cordial relations with competitors' that Martin identified is nowhere to be found in current relations between the communities.

The origins of the Renews' Rock controversy go back many decades, when the Newfoundland government (or, as Martin would put it, 'the law in St John's') passed a regulation governing the type of gear that could be used in the Renews' Rock area. Most Fermeuse fishermen cannot remember how long the law has been in existence. Most simply said, 'It's been going on as long as I can remember.' One respondent, however, was able to identify the regulation as dating from 1927 (Respondent no. 106). The regulation limited fishing in the Renews' Rock area to handlining only:

The law was made up years ago to only allow people to go handlining and jigging. (Respondent no. 104)

At the time the law was instituted, it was clearly designed to eliminate the use of trawls and cod traps in this relatively small area:

There's a law on the books that you're not allowed to trawl here [on Renews' Rock]. (Respondent no. 100)

The dispute is between handliners and those who want to set trawl, and that has been [going on] right through the years. (Respondent no. 100)

Today, however, it also functions as a prohibition against gill-netting:[8]

The fishery was different then. Gill nets weren't available, and trawls were available on a limited basis, and the majority of fishermen handlined. (Respondent no. 106)

You're not allowed to trawl or gill net up there from June 1 to the last of October. (Respondent no. 115)

Just why the Newfoundland government introduced this law in 1927 is not known. One respondent pointed out that Renews' Rock is probably the only small local area in Newfoundland where the use of any fishing technology other than handline and jigger has been prohibited:

It is the only area I know of that is a closed area designated to a certain type of fishing, and I think that this is discriminatory. I feel that you should've been able to gill-net and trawl there. (Respondent no. 106)

A probable rationale for the restrictions is, however, readily apparent. The Renews' Rock area has always been a prime area for cod fishing. Fermeuse fishermen described it as an area that is 'black under the water with fish' (Respondent no. 104). Given the small size of the area and the density of fish in it, the use of a more sophisticated technology would cause the resource to be depleted by just a few fishermen in a very short time. By limiting the technology to be used in the area to handlines, the government ensured that a substantially larger number of fishermen would be able to share this rather special resource.

It is worth emphasizing that the Renews' Rock law did not prohibit Fermeuse fishermen from fishing there; it prohibited them only from using trawls, traps, or gill nets in the area. Most Fermeuse fishermen were vehement in their objections none the less; they felt they should have the right to use trawls in the area:

We should have the right to trawl up on Renews' Rock ... We can go handlining up there too, but that is no good to us. We would only make a couple of hundred dollars a week. (Respondent no. 098)

You go up to Renews' Rock with thirty lines of trawl, and you come back with 3000 pounds [of cod]. But you can't get it handlining. A lot of fellas [from Fermeuse] went up yesterday [to go handlining] – three fellas in a boat – and got only 180 pounds. Now that's nothing! (Respondent no. 110)

It is also quite clear that, in the past, Fermeuse fishermen respected the Renews' Rock prohibition more in the breach than in adherence. Many of those we interviewed spoke of having gone trawling in the Renews' Rock area despite the legal prohibition:

You wouldn't know there was fish there only for the trawls. My sons and their cousin were trawling up there the [i.e., this] year and cleared $1700 each in one week. (Respondent no. 109)

Much of this illegal fishing was done at night, and some of the transgressors were clearly proud of their nocturnal accomplishments:

We used to have a go up there in the night-time to get clear of the patrol boat. (Respondent no. 114)

We used to fish all night and come in at five o'clock in the morning. (Respondent no. 104)

Up to last week we used the radar to go off the Rock at night – 10 [P.M.] to 3 [A.M.] ... and it was really good. No one ever used radars around here until three years ago. (Respondent no. 114)

A couple of fishermen went so far as to assert that it was the bait on the trawls of Fermeuse fishermen that kept the fish in the Renews' Rock area:

Last year we were trawling and we were keeping the fish there with our bait. (Respondent no. 114)

I can't see why you can't trawl up there. There are about two hundred men who would like to trawl up there. I feel the nets do put men out of business, especially fellas with traps. But I don't see anything wrong with lines of trawl. I feel that trawl even brings fish in to the handliners. They're getting no fish up there handlining. I can't see how twenty-five to thirty men can keep two hundred trawlmen from fishing. (Respondent no. 115)

Throughout this book, we have argued that the fundamental divisions in the inshore fishery are those that exist between users of different technologies, and between full- and part-time fishermen. At least as the Fermeuse fishermen tell it, both these divisions underlie their conflict with Renews fishermen. Fermeuse fishermen have long wanted to use trawls in the Renews' Rock area. Renews fishermen have just as steadfastly opposed the practice.

It's been going on for as long as I can remember. The Renews crowd wants it for handlining, and the crowd over here wants it for trawls. (Respondent no. 102)

The fishermen from the Renews area want only to handline in the Renews' Rock area, whereas the fishermen from the Fermeuse area want to gill-net and trawl there. However, the handlining is protected by law. (Respondent no. 097)

It should be noted that Fermeuse fishermen see the situation as involving not gear conflict, primarily, but rather a conflict between full-time and part-time fishermen. They forcefully assert that the source of the intercommunity conflict is the fact that the majority of Renews fishermen are part-timers who have other jobs, whereas the Fermeuse fishermen are mostly full-time fishermen who have no other employment.[9] They argue that the law in effect protects part-time fishermen rather than the 'bona fide fishermen' they consider themselves to be. It is this factor that seems most to arouse the ire of Fermeuse fishermen and that infuses the situation with so much tension:

We should have the rights to trawl up on Renews' Rock, 'cause most of 'em handlining up there are part-time fishermen. (Respondent no. 098)

A lot of fellas from Renews – about 50 per cent – are against us trawling and gill-netting there. Part-time fellas should have no say about restricted areas because he is not depending on it for a living. (Respondent no. 105)

You could count the fishermen [at Renews' Rock], and you're likely to get two-thirds part-time fishermen. We've got terrible overcrowding. (Respondent no. 100)

Well, up at Renews' Rock, this is our trade, but other people who have other jobs are up there too. But this is our business. It doesn't matter to other people if they don't get fish, but it matters to us. It's our business ... Just about all the people in Fermeuse are full-time fishermen and just about all the ones in Renews are not. (Respondent no. 112)

The Renews' Rock controversy has not only divided Fermeuse and Renews fishermen, it has also driven a wedge between two groups of Fermeuse fishermen – men who share the same general harbour and fish the same grounds. Moreover, the intracommunity conflict is most assuredly over gear and not over part-time versus full-time fishing status. The matter came to a head when one of the fishermen of Fermeuse solicited signatures on a petition requesting that the Department of Fisheries open up the Renews' Rock area to trawling. Some of the fishermen refused to sign. The resulting charges and countercharges also revealed a wide range of generally questionable fishing practices in the Renews' Rock area:

They're going around with a petition to change the law to allow trawling up there. Some people has gill nets and they won't sign the petition because they wouldn't want to take up their gill nets to allow trawls to go in. They trawl now, but during the trap season they set out gill nets even though they're not supposed to. (Respondent no. 104)

Gill nets set close to the cod traps are turning the fish (away) from the traps. They set their gill nets between the Rock and the traps, see. (Respondent no. 107)

It is clear that, even though the Renews' Rock area was given its special status as early as 1927, major conflicts about it did not occur until quite recently:

We were there twenty-five years and we had no trouble until this year. I think a bigger crowd of fellas with trawls were at it this year. (Respondent no. 109)

It is therefore relevant to consider what has occurred to change the carefully crafted equilibrium that Kent Martin identified as a fundamental feature of the area's social structure in the early 1970s. We would suggest that there are two categories of causes behind the demise of traditional structures for maintaining social harmony.

The more immediate causes are the recent actions of Fermeuse and

Renews fishermen themselves. We have already noted that a petition was circulated in Fermeuse requesting that the government lift the ban on trawling in the Renews' Rock region. The gathering of signatures was still under way during our time in Fermeuse:

We petitioned thirty-eight [people] today, thirty-five for and three against, and there is a person in Renews trying, too. But I don't think you are going to do well there, 'cause they mostly handline. Only full-time fishermen are allowed to sign the petition. (Respondent no. 098)

But petitioning was only part of the political offensive mounted by the Fermeuse fishermen. Those we interviewed also spoke of a series of meetings between Fermeuse fishermen and government representatives to discuss the situation:

There are people, our committee here, started last July to change it. We had eighteen meetings with the Department of Fisheries in order to get the gill net and trawls changed. (Respondent no. 097)

In response to the Fermeuse offensive, the fishermen of Renews took action of their own. They complained often and vehemently to the Department of Fisheries about illegal fishing activities of Fermeuse fishermen in the Renews' Rock area. The Department of Fisheries first responded by sending an occasional patrol boat to the area. But that was not enough to stop the illegal fishing or to halt the Renews fishermen's vociferous complaints. Consequently, by the time we were conducting interviews in Fermeuse, two Fisheries patrol boats had been assigned to monitor the small Renews' Rock area twenty-four hours a day:

The only reason I'm talking to you now is the fact that patrol boats are around the Renews' Rock area. (Respondent no. 115)

The Renews crowd phones up the Fisheries Department and keeps the (patrol) boats out all the time ... This year, boy, there's great fishing, but we're not allowed to get it – and in a week or so it will be gone. (Respondent no. 104)

We can go handlining up there but that is no good to us. We would only make a couple of hundred dollars a week. Now they got two patrol boats up there the past week, instead of one, and they have got a twenty-four-hour watch on. (Respondent no. 098)

We used to have to go up there in the night-time to get clear of the patrol boat. But they put it on twenty-four hours. (Respondent no. 114)

We used to fish all night and come in at five o'clock in the morning. But then they started sending the patrol boat out at night. (Respondent no. 104)

Although the actions of Fermeuse and Renews fishermen that resulted in increased government patrolling of the Renews' Rock area were quite clearly the immediate cause of intercommunity hostilities, our interviews gave considerable evidence that they were in fact symptoms of more deep-seated problems. We found that the number of fishermen, both full-time and part-time, in the Fermeuse–Renews area had increased dramatically in recent years, while the available stock of fish had simultaneously declined drastically. At the time that we were there, many fishermen were lamenting their inability to capture any fish by what had traditionally been the most reliable method, the cod traps. Indeed, some were predicting that several crews of fishermen would be forced to give up trap fishing after the current season:

There are fellas who will be turning in their berth next year because there is no fish and fellas can't afford to be at it any more. Well, for us fellas, if you could get 80,000 to 100,000 pounds in the traps, it would be worth while. There are about five fellas who will drop out next year. There's no fish in any of the berths this year. (Respondent no. 105)

I say in three years, all the traps will be gone here. The fella I'm fishing with says this will be his last year. He's selling the whole thing and getting a speedboat to go jigging, trawling, and handlining. (Respondent no. 104)

In view of the competition for trap berths that had long prevailed and the possessiveness with which crews held on to their berths, such predictions were dire indeed. Faced with the conditions we have described, Fermeuse fishermen were clearly approaching a stage of desperation. There was no fish in the Fermeuse area generally, and certainly none was available through trap fishing. By contrast, 'the sea was black with fish' in the Renews' Rock area. It was controlled desperation that fuelled their mission to remove the proscription against trawling and that caused them to regard the Renews fishermen as, primarily, part-timers who were appropriating the only basis of livelihood left available to them. The following comments express their state of mind eloquently:

We ought to have some kind of chance. That's all we got [i.e., Renews' Rock]. (Respondent no. 114)

One man had his trawls on the Rock and they were taken by the patrol boat. And the next day he come on board the patrol boat at the wharf and went into the pilot house and told the skipper, 'If you take my trawls again, I'll drown you in the water and I'll drown myself to drown you. Take away my trawls and you're taking away my reason for living.' And he is one of the quietest men around here – he just lost his temper. (Respondent no. 104)

'The future of the fishery is the fish plant.'
(Respondent no. 097)

As in Grates Cove, so in Fermeuse, the residents were unanimous in liking where they lived. Indeed, as Table 6.2 shows, there was positive unanimity on four of the five community-evaluation measures used in the study. In the words of one respondent, 'I love to live here' (Respondent no. 107); another declared that he 'wouldn't move for the world' (Respondent no 100). The only reservation appears to be some hesitancy in a few people's minds about the quality of community leadership.

For the most part, Fermeuse residents were also satisfied with the level and quality of community services. As in Grates Cove, they had high praise for the community's schools and teachers, and most were satisfied with the level and quality of medical care they received. Still, like Grates Cove residents, they were almost evenly divided on whether there was enough entertainment to satisfy people, and on whether or not they had to do without a lot of the things that those living elsewhere might have.

Once again, we see this positive evaluation of a community dissipate quickly when we examine responses about the future of the local fishery and employment opportunities in the community. Only one respondent thought of Fermeuse as a good place to find work; only two thought the fishery there had any future; and only two thought the inshore fishery offered good opportunities for young people.

TABLE 6.2
Fermeuse respondents' views of their community and its future

	Yes	No	Uncertain
Evaluation of community			
Like living here	20	0	0
Good place to raise children	20	0	0
People here generally hard-working	20	0	0
Leaders are good, capable people	15	1	3
People here care about the place	20	0	0
Evaluation of community services			
Good schools and teachers	20	0	0
Good medical care	17	2	1
Enough to do in spare time	12	8	0
People here do without a lot	10	8	2
Evaluation of employment opportunities			
Good place to find work	1	19	0
Fishery has a future here	2	12	6
Fish plant here has a future	10	2	8
Inshore fishery good for young people	2	17	1
Evaluation of community's future			
Children should settle here	11	2	1
Community has a good future	9	3	8

Note: There were 20 respondents. Where responses for any given item do not total 20, the balance represents abstentions.

A pivotal factor in this assessment of the fishery was the perceived future of the community fish plant. The plant had been closed some weeks prior to our arrival in Fermeuse. The hopes of many residents lay in the rumours circulating in the community (and on the regional radio stations) that the plant would soon be taken over by another fish-processing chain. Consequently, the respondents were evenly divided between those who thought that the local plant had a future and those who were uncertain or without hope:

It just has to turn around. I think it has to. (Respondent no. 112)

This plant is close to the Northern cod. It has a lot of advantages. (Respondent no. 115)

This plant is going to have to change if the fishery is going on – poor management. There are too many white hats walking around down there. (Respondent no. 098)

It's hard to say. Now if you can get another company to take over the plant, it'll be OK. (Respondent no. 103)

All our respondents agreed with the view expressed in the quotation that starts this section, namely that, unless the plant reopened, the future of both the community and its fishery were dim. Many of those we interviewed saw little opportunity for themselves outside the fishery:

What do I do then if I give it up? (Respondent no. 102)

They also saw little future for any children who might think of following in their parents' footsteps:

[I'd tell them to] stay clear of it. It costs too much money to go into it. (Respondent no. 097)

No, [don't go into the fishery]. Not if he can get another job. (Respondent no. 113)

The following comments reflect the deep pessimism of Fermeuse residents about the future of their community:

It's only going downhill. The plant close down and then wait for the government to give them money to open up. (Respondent no. 101)

It used to be a good occupation. But not any more. (Respondent no. 112)

Conclusion

This chapter has cast the issues surrounding community versus state control of common property in a somewhat different light from that of the last chapter. The two communities discussed there provided examples of distinctive forms of community control of common property both prior to and in combination with recent changes in state regulation. In both Charleston and King's Cove, community control was vested in traditional and largely implicit measures. In contrast, in both Grates Cove and Fermeuse, community control is manifested at a far more formal and more explicit level, to the point that in Fermeuse, in particular, local control comes into direct and often open conflict with state regulation.

Furthermore, in both Grates Cove and Fermeuse, there was a close interdependence between the nature of community regulatory practices with regard to fishing property and the various ramifications of four dominant issues in the inshore fishery, namely, (1) the relationship between full- and part-time fishermen; (2) the relationship between those who hold protected-species licences and those who do not; (3) the relationships among those who use different modes and different technologies of production; and (4) the relationship between control of common property and the unemployment-insurance system, which provides some modicum of income security.

An examination of the interrelationship of state and community regulations, combined with actual community practices, also affords a clearer understanding of the social relationships and divisions within the inshore fishery. Thus, state and community practices with regard to both the access to and the process of fishing have been shown here to create many of the social and economic class divisions that exist among fishermen. Put somewhat differently, a fisherman's class position within the inshore fishery is not determined by his relationship to the means of production alone; instead, as demonstrated by numerous instances described here, state and community regulations governing the access to and the process of fishing are shown to have a significant influence on a fisherman's economic opportunity and, ultimately, on his social class position within the community.

7 A House Divided

Competition for Scarce Resources in
Newfoundland's Largest Inshore
Fishing Centre

The fishery is sort of a house divided amongst itself – gill nets, traps, longliners.

A Bonavista fisherman
(Respondent no. 050)

There's three kinds of fishermen in Bonavista – inshore fishermen, longliner fishermen, and crab fishermen. And there is three kinds of boats and they're all the time stabbing each other in the back. The inshore, the small boats, are down on the longliner fishermen. I don't like it. It's not right. Let every man, if he's got guts enough to do it, let him go ahead and get what kind of boat he wants. And that's why the union is no good in Bonavista – it's got one fella penalizing the other.

A Bonavista crab-boat skipper
(Respondent no. 070)

BONAVISTA

Bonavista is the largest inshore fishery community in Newfoundland, if not in the world. The 1986 census lists its population as 4605, making it the fourteenth-largest community in Newfoundland and Labrador. However, each of the thirteen larger communities in the province has a major source of employment other than or in addition to the fishery, or is within easy commuting distance of St John's, which provides an alternative source of employment for much of its labour force. Of greater significance for our purposes is that Bonavista has more licensed inshore fishermen than any other community in the province. Federal fisheries licensing records show that there were 232 licensed full-time fishermen and an additional 198 licensed part-time fishermen living in Bonavista at the time of our study. Finally, it should be emphasized that Bonavista is the only fishing port of major size in Newfoundland that depends solely on the inshore fishery; that is, no vessel larger than a longliner operates from Bonavista, even though the community has a large fish plant and a major crab-processing plant. Bonavista can therefore be viewed as a perfect example of a major inshore fishing centre, and it was on that basis that we chose to include it in this study. Compared with the other communities examined here, Bonavista is clearly in a class of its own, both in terms of size and the complexity of its fishing practices.

Like King's Cove and Grates Cove, Bonavista is one of the oldest communities in Newfoundland. According to local lore, it was the location of John Cabot's 1497 landfall in North America, and its name derives from his declaration 'Oh, beautiful sight!' in his native Italian. To affirm that apocryphal claim, the Newfoundland government has erected, on nearby Cape Bonavista, a huge bronze statue of Cabot surveying his 'newe founde land.' A tourist brochure put out by the Town Council of Bonavista describes Bonavista's history in the terms in which the community sees itself and wishes outsiders to see it:

Bonavista, one of the oldest towns in North America, is situated on the East Coast of Newfoundland about 300 kilometres by road North of St John's. Cape Bonavista, located in the Northeast part of the town, was the point of land first sighted by John Cabot, discoverer of Newfoundland, on the morning of the 24th of June, 1497. The town of Bonavista was first settled towards the end of the sixteenth century by the English, who came to fish.

Between 1696 and 1705, the French attacked the town on four occasions and in 1705 the inhabitants took refuge on Green Island, just offshore from Cape Bonavista, where it is said cannons can still be seen under water. In 1722, a Church of England clergyman, the Reverend Henry Jones, built the first church in the town and also the first school in Newfoundland. In 1760, Captain Cook visited Bonavista on one of his survey journeys and in 1842, the fourth lighthouse in Newfoundland was erected at the Cape.

Certainly, through the eyes of a fisherman, Bonavista is a 'beautiful sight' indeed. It is located just down from Cape Bonavista (the tip of the Bonavista Peninsula, which separates Bonavista Bay from Trinity Bay), and thus occupies much the same position on that point of land that Grates Cove occupies on the other side of Trinity Bay. Indeed, Bonavista is only about thirty kilometres by boat from Grates Cove, directly across Trinity Bay, though it is approximately four hundred kilometres distant by road. Bonavista's headland location, like Grates Cove's, not only gives its fishermen close access to the bounteous fishing grounds off 'the Cape,' but also puts them within easy reach of the fishing grounds on both the east side of Bonavista Bay and the west side of Trinity Bay.

The additional attraction of Bonavista is that it has a large and reasonably well protected natural harbour, surrounded by land that is generally flat (at least by Newfoundland standards), well watered by natural streams, and reasonably arable (again, by Newfoundland standards).Thus, unlike Grates Cove, where the terrain presents at least as great a challenge as the sea, Bonavista provides virtually perfect conditions for a large inshore fishing community.

Like other communities on Newfoundland's northeast coast, Bonavista was settled early by fishermen from the south of England. Most such settlements, however, were inhabited only during the summer fishing session, with a handful of 'men servants' staying over during the winter to tend to the gear and sheds that were left behind for the following season. Such communities had few women and children, and most of the latter were boys who had accompanied their fathers as helpers in the fishery (Macpherson 1977, 104). Still, Bonavista was settled permanently earlier than most other areas of Newfoundland: the List of Planters' Names for 1676 describes Port Bonavista as having been 'settled on or before the year 1660' (quoted in Macpherson 1977, 105).

One of the more interesting ways of establishing the extent of early permanent settlement in a community is to search out the earliest record of the family names of its current residents. We did this in the case of Bonavista with the aid of the list of fisheries licence holders in conjunction with Seary's *Family Names of the Island of Newfoundland* (1977). Our research revealed that the forebears of many of the community's current residents were settled in Bonavista well before 1800. The earliest record is of George Sciffington, 'chief Quaker of Bonavista,' in 1705, and of William Kean, 'merchant of Bonavista about 1719.' These settlers were subsequently joined by early representatives of the Abbott (1765) and Ryan (1774) families. Indeed, in the case of the latter, the record names a woman, Margaret Ryan, which suggests that entire families were residing in Bonavista by this time. There are records of the Blackmore and Rolles families dating from 1781 and

1784, respectively. Written records, however, were rare, so the presence of other families before the 1780s cannot be verified. With the establishment of a Church of England register for Bonavista in 1786 (Macpherson 1977, 102), record keeping improved. Thus, we can observe, in the years immediately following, the arrival of the forebears of many more of today's residents.[1] It is clear from all this that, more than two hundred years ago, Bonavista was a thriving fishing centre.

Unlike the smaller fishing communities that we have examined thus far, Bonavista is a sprawling conglomerate of century-old wooden houses and modern bungalows situated along a veritable maze of narrow, twisting lanes that wind around the harbour and back over the nearby hills. Although few of those lanes have street signs, many have names that are well known to the local residents. Bonavista is a town composed of distinct neighbourhoods that were originally separate communities and that are still known to local residents by their original names – for example, Bayley's Cove, Canaille, Mockbegger, and Red Point. Those names do not appear on tourist maps, but they are the basis on which local residents identify where they live and give directions to strangers visiting the community.

The size of the fishery labour force in Bonavista and the diversity of the fish stocks in the area produce a wide range of licensing arrangements. Ninety of the 431 licensed fishermen in the community hold a total of 130 protected-species licences. Their distribution is presented in Table 7.1. The most common are salmon licences, which are held by 50 full-time and 12 part-time fishermen. Next, there are 43 herring licences, all held by full-time fishermen, followed by 11 lobster licences, held by 10 full-time fishermen and 1 part-time fisherman. There are also 5 capelin licences and 1 otter-trawl licence. However, as we will see shortly, the most significant species licences are the 6 crab licences, held by 5 full-time fishermen and 1 part-timer. Crab licences have the potential to make 'highliners' out of those who possess them.

One indicator of the degree of a community's involvement in the fishery is the ratio of full-time to part-time fishermen in its fishery labour force. Even in Grates Cove, which we described as a decidedly fishing-oriented community, that ratio was only 1 in 4.8 – that is, only 1 out of every 4.8 licensed fishermen held a full-time licence. The implication of this is that, even though many people in the community might fish for a portion of their living, only a relatively small number devote their entire economic activity to it. In Fermeuse, where the conflict between fishermen who used different technologies of production was

TABLE 7.1
Distribution of species licences among Bonavista fishermen,
by general fishing-licence status

Licence combination	Full Time	Part Time
General fishing licence only	157	184
General fishing licence and salmon licence	25	12
General fishing licence and herring licence	16	–
General fishing licence and salmon and herring licences	15	–
General fishing licence and lobster, salmon, and herring licences	5	–
General fishing licence and crab licence	4	–
General fishing licence and lobster and herring licences	3	–
General fishing licence and lobster and salmon licences	2	–
General fishing licence and herring and capelin licences	2	–
General fishing licence and salmon, herring, and capelin licences	2	–
General fishing licence and lobster licence	–	1
General fishing licence and salmon and crab licences	1	–
General fishing licence and crab licence and otter trawl	–	1
General fishing licence and capelin licence	1	–
Total	**233**	**198**

intense, the proportion of full-time to part-time fishermen was even smaller, at approximately 1 in 10. Hence, it is a strong indicator of the commitment of Bonavista to the fishery – and of its dependency on it – that better than 1 out of every 2 fishermen (specifically, 54 per cent) in that community holds a full-time fishing licence. Not only are there far more fishermen in Bonavista than in any other community we have considered thus far, but a far higher proportion of them depend almost entirely on the fishery for their livelihood. Primarily because of the size and diversity of the inshore fishery in Bonavista, we interviewed a much larger sample of fishermen there than in any of the other communities we studied. In all, we conducted interviews with fifty fishermen. Of those, thirty-nine were full-time fishermen and seven were part-time fishermen. An additional four people whom we interviewed had held part-time licences during the previous season (and were therefore included in the licensing lists) but had not renewed their licence in the current year.

As in the other communities we studied, we selected the sample to be interviewed by dividing the fishermen into full- and part-time licence holders, then drawing a random sample from each group.[2] Because it was clear to us by now that part-time fishermen offered less insight into the nature of fishing practices than did full-time fishermen, we deliberately oversampled full-time fishermen and included only a small random sample of part-timers. Finally, our random sample of full-time fishermen happened to contain two captains of crab-fishing boats. From their comments as well as from our other interviews, it became apparent that the crab fishery was unique and important, and that it deserved special attention. We therefore interviewed an additional five crab-boat captains, even though they did not fall into our random sample. (We included those in the total of thirty-nine full-time fishermen interviewed.)

'You got to have the how with it.' (Respondent no. 048)

As noted in previous chapters, the fishery is essentially a hunting activity in which the participants are in fundamental competition for the same limited resource. Indeed, that competition is the basis of the 'tragedy of the commons' perspective's depiction of the fishery as inherently involving the unregulated pursuit of an open-access resource. In view of the large number of fishermen operating out of one harbour in Bonavista, the level of competition in this community is likely to be high: No matter how good the harbour may be, the fishing grounds adjacent to it are always limited.

As we have seen, although competition in the fishery may take a number of forms, it inevitably resides in a conflict between different types of boats (means of production) and different types of gear (technologies of production). In some places, such as Fermeuse, similar boats using different types of gear were in competition against one another, while in others, such as Grates Cove, both the types of boats and the types of gear differed. As might be expected, the Bonavista fishery is sufficiently complex that it embraces virtually the complete range of boat and gear combinations.

Traditionally, the cod fishery around Bonavista was a handline (that is, jigger), trawl, and cod-trap fishery involving relatively small boats. Under such conditions, the advantage, where there was one, usually went to those who had the best understanding of the fishing grounds. Consequently, in Bonavista, old-timers speak with pride of their skill and knowledge in this regard:

On me ninth birthday they took me out [i.e., to begin to fish for a living]. There was not engines, only sails. I made $88 that year. Me father put it in the bank. I got the old bankbook there now. I fished all me life with a piece of wire crooked up [i.e., the bent hook on a jigger], and trusted the fish to swim into it. And I never had a cent of government [i.e., welfare] ...

There was always good fishermen and always will be. People got larger boats to [i.e., than] me brother and myself and can't make a success of it. You got to have the how with it. They can't seem to learn, eh. If you're trawling you got to know the [fishing] ground, and if you're handlining you got to know the ground, eh. Lots of cases the bad ones [i.e., poor fishermen] tries to get the big boats. (Respondent no. 048)

I knows the ground now just like you knows a fence around your house. (Respondent no. 055)

However, as in the other communities we examined, the essential nature of the fishery in Bonavista was forever altered by the introduction in the 1970s of a new type of gear and a new type of boat, namely, the gill net and the longliner.

'With nets, one may can spoil a place
where twenty men can drop.'
(Respondent no. 049)

The gill net was introduced into the Newfoundland fishery in the 1970s as a direct policy of the provincial government designed to increase the incomes of fishermen. Coupled with government incentives to encourage fishermen to build longliners, the gill-net technology was seen as a way of establishing a core of financially secure fishermen who could make an adequate living from the sea. However, as is clear from our interviews in Bonavista, many of the traditional, small-boat fishermen saw the introduction of gill nets as a 'curse':

The biggest curse in Newfoundland is when they sent [a government Fisheries official] to Iceland and he bring back the gill nets. (Respondent no. 049)

They argued vehemently that the use of gill nets in the inshore fishery threatened not only to limit the access of other fishermen to their traditional fishing grounds, but also endangered the survival of the fish stocks themselves.

Gill nets are usually set in the same shoal water areas that are prime sites for handlines and traps. This makes access problematic for handliners and trap fishermen in two ways. First, handliners allow their boats to drift without power over the shoals, and the presence of gill nets greatly increases the likelihood that their hooks and lines will become tangled:

I feel that the gill nets should not be allowed inshore. Because a man with a hook and a line gets tangled up 'cause if he is on [i.e., above] a rock he got to move too close to the gill net. (Respondent no. 026)

The second, and far greater, problem for these handline and trap fishermen is that a 'fleet' of gill nets set in their traditional fishing places will not only capture many of the fish that would otherwise be theirs but also make it virtually impossible for them to reach their prime fishing spots.

The gill net should be done away with. By the Lord Jesus they should put a ban on it. The first thing ever since I was twelve years old you could go down to Gull Island off the Cape there and you could see four to five boats 'trapped' in the rocks where you wants to catch your fish. Now you can't get near to it because of the gill nets. (Respondent no. 034)

The gill nets are nothing but a frigging nuisance. Now you go on a rock, you can't go there because they are put into five-fathom water, and that blocks things off. From Blackhead down, there is only two or three places you can fish because of the gill nets. They shouldn't be allowed to put 'em out inshore ... cause we are losing fish. We got to do something. (Respondent no. 055)

We have already seen that handline, trawl, and cod-trap fishermen in other communities are convinced that gill-net fish is inferior in quality to fish caught by the more traditional methods. Bonavista fishermen are no exception. They provide graphic descriptions of the poor quality of fish left to 'rot' for days in gill nets:

It breaks your heart to see a fellow come in with fish that has slub [i.e., scum] on it and the eyes eat out of it in the gill net. And he is getting twenty-seven cents a pound. (Respondent no. 048)

Where we fish the past week, a fellow had a gill net set on Friday and hauled it yesterday [i.e., Tuesday]. That's five days with fish there. There's fish that is rotten coming out of gill nets. (Respondent no. 48)

To further emphasize the legitimacy of their position, Bonavista fishermen provide vivid descriptions of gill-net fishermen who, when they want a fish for their own table, trade one of their gill-net fish for one caught by hook and line or by trap:

The 'codhoppers' I call 'em. They want a cod fish to eat, they'll come over and trade him. If they won't eat 'em, why should you? They puts 'em right on your table. (Respondent no. 055)

There's not one man in Bonavista 'll eat 'em. He'll come to a man with a handline and change it [i.e., for his own consumption, a gill-net fisherman will trade his fish for one caught by a handline]. (Respondent no. 037)

Finally, to add insult to injury, as in other communities, the union has negotiated a pricing system with the fish plant that favours the gill-net fish over that caught by handline, trawl, or trap:

A man with a cod trap here in Bonavista, he goes out in the morning and within one hour he's hauled in [i.e., got his fish in his boat] and two hours later he's getting in [to port]. That fish is alive on the [splitting] table. He only gets [a few cents] a pound. He gets less than gill nets. (Respondent no. 047)

You can get [more money] with gill-net fish here, and he'll smell [i.e., be rotten]. (Respondent no. 037)

Few of the handline, trawl, and trap fishermen whom we have interviewed spoke well of longliner fishermen who fished the same inshore grounds as they did. In the inimitable words of one such fisherman,

[A local fisherman] has the place smothered with gill nets. You've got to have a brass face to do what he do. He got that longliner, see. It makes me sick to see all them gill nets, and in ten fathom of water, too. In shoal water. (Respondent no. 048)

However, the pressure on inshore fishermen comes not only from longliners using gill nets. They also see themselves as victimized by other types of fishermen and gear. Notable among those are the herring seiners. As noted above, forty-three Bonavista fishermen held herring licences in some combination with other licences. Despite having to pay an annual fee to retain these licences, in the three years prior to our interviews with them, they had been unable to fish for herring at all. This was because herring seiners had already taken up the low quotas of herring permitted to be caught annually in Bonavista Bay. (Herring generally enter Bonavista Bay from the west side, where the seiners operate, so by the time they reach the waters on the east side, around Bonavista, the allowable quota has already been filled.) The Bonavista fishermen's frustration with this practice is captured in the following comment:

What happens to we is the seiners. The seiners takes up the quota. We got the licence because if you got 'em you can keep 'em. Otherwise we'd lose 'em. (Respondent no. 055)

Finally, the small-boat fishermen of Bonavista point to their fundamental conflict with large offshore draggers. These ocean-going vessels 'drag' huge nets over the offshore fishing grounds, scooping up all species and sizes of fish. This indiscriminate action is seen by Bonavista fishermen as problematic because they believe not only that it captures all the fish stocks that would otherwise grow and come inshore, but also that it destroys the marine plant life on which those fish feed. In the words of one, dragging 'takes the flowers off the rocks':

They smothered us. They ruined the ground, see. They takes the flowers off the rocks the fish eats. (Respondent no. 055)

Draggers, gill-netters, longliners – they fishes all year round and that destroys fish. They goes dragging where the fish feeds, and that is all dragged to pieces.

Too much dragging winter and summer – they should stop for a couple of months to let the fish spawn. (Respondent no. 032)

You take these draggers. There's just as much killed on the bottom with these iron bars [used to hold the nets on the bottom] as is brought in. There'll soon be ne'er one [neither one] left. Now longliners are not overfishing. They're only fishing in summer same as small boats. (Respondent no. 038)

The federal government should take a licence away from two foreign draggers and give it to 150 fishermen. Then we'd have a better province to live in. (Respondent no. 041)

In sum, the small-boat inshore fishermen see themselves as being at the bottom of the hierarchies of both the inshore and the offshore fisheries, and, hence, as being in a most vulnerable position. They consider that vulnerability to be a direct result of conflicts among the different technologies used to capture fish. However, they also believe that it is attributable to government regulations and fishery-union actions that, they allege, have discriminated against them in terms of both their right to catch fish and their right to sell fish. Their views on the subject of government regulation will be the focus of a later section of the chapter. Our discussion here, however, provides further validation of our earlier contention that fishermen are in competition and potential conflict over both the right of access to the fishing grounds and the right of access to markets for their fish. Any regulation, whether it is issued by the provincial or federal government, the union, or any other agency, that affects either of these two spheres of competition has the potential to alter drastically the nature of fishery work in Newfoundland's rural communities.

'But the gill-netters don't have much of a chance.'
(Respondent no. 062)

One possible response to the complaints of Bonavista small-boat fishermen is that they are simply victims of 'progress.' After all, gill nets and longliners were introduced largely through government grants and subsidies aimed at moving a segment of inshore fishermen from a 'marginal work-world occupation' in which thousands of small-boat fishermen made an inadequate living and the fishery was seen as the 'employer of last resort,' to a 'central work-world occupation' in which the fishery would become a 'profession' for a core of adequately renumerated workers. In this scenario, a large proportion of traditional inshore fishermen operating from small, unproductive boats must necessarily be sacrificed to ensure that the fishery becomes a viable occupation for a forward-looking body of longliner fishermen. Judging by the rates for fish that it negotiated with the fish plants, it would appear that the fisheries union supports this model: longliner fishermen are assured of a higher rate of pay for gill-net fish than small-boat fishermen receive for handline and trap fish. However, our interview data reveal a very different image of longliner fishermen from the one that emerges from this set of assumptions. Far from being successful 'captains of their fate,' they too are the victims of a set of regulations and policies that seem destined to reduce the majority of them to bankruptcy. Although it is probably not true that, as one small-boat fisherman stated, 'the bad ones tries to get the big boats,' most longliner owners are nevertheless deeply in debt, largely as a result of government policies that encouraged them to switch to the bigger vessels. The extent of such debt is reflected in the following descriptions from two largely unsuccessful longliner owners:

I got a new longliner, only a couple of years old. It cost $400,000. The [Fishery] Loans Board gave [i.e., lent] me $200,000. The provincial government gave me 15 per cent and the federal government gave me 25 per cent. I had to make a down payment of $20,000. I had my own longliner before that – a smaller one – and that's where I got the $20,000. (Respondent no. 058)

My boat cost $430,000. The federal subsidy is 35 per cent and the provincial subsidy is 15 per cent. The rest, the government puts up as a loan. (Respondent no. 046)

In short, for a relatively modest down payment, a fisherman can purchase a sizeable vessel and several hundred thousand dollars worth of debt.

Of course, boats of this size also require a considerable sum to operate. Many longliner owners we interviewed estimated that their annual fuel bill alone amounted to close to $20,000, while insurance could approach $1,000 per month.[3] In addition, there were gill nets to be bought, often with a loan from the fish plant as an advance on catch earnings.

The servicing of such debts costs money. As another longliner owner noted, 'A $400,000 boat – you got to get a lot of fish for that' (Respondent no. 048). Although the Fishery Loans Board may give loans at interest rates considerably lower than prevailing bank rates, it then turns them over to regular banks for collection. The Board pays the difference between the loan rate it offers to fishermen and the current loan rate available from the bank. As long as the fishery is good and the owner remains in good financial standing with the bank, this arrangement poses no problems. But, in a bad season, the longliner owner may not be able to pay the thousands of dollars needed each month to service his debts and operating expenses. A national bank, with its head office in Toronto or Montreal, unfamiliar with the annual exigencies and unpredictability of the Newfoundland inshore fishery, is not likely to waive payments in a bad year. A longliner operator suffering from poor catches may thus find himself the victim of foreclosure on his outstanding loans as well:

I got to come up with $27,000 this year for payment: $20,000 interest and $7000 payment – where am I going to get it? (Respondent no. 046)

The consequence is that, in Bonavista, as all over Newfoundland, there are longliners tied to the wharfs or left abandoned on beaches. Some have been confiscated and thus become the property of banks, which offer them for sale at greatly reduced prices. But few fishermen are willing to buy them, having seen at close hand the consequences of such purchases. Other longliners remain the property of their original owners, who have given up operating them, as they have learned that the cost of doing so exceeds the income the boats can produce.

Two of the longliner owners we interviewed in Bonavista had abandoned their boats and were back fishing out of trap boats with their fathers and brothers. In doing so, they were following the example of many former longliner crew members who had come to learn the hard way that they could not obtain an adequate living as a shareman on a longliner:

There's a movement from longliners to trapboats. (Respondent no. 050)

Indeed, it would appear that the only way to make longliner operations profitable is through possession of protected-species licences. As we shall see in the next section, the most valued of these is the crab license, but, if that is not available, then it is absolutely mandatory to have a salmon licence and probably lobster and herring licences as well. Without these, and without the opportunity to use them, it is virtually impossible to make a longliner profitable. One longliner owner who has been denied crab and salmon licences expresses his frustration this way:

I've got a full-time licence and I'm not even allowed to fish. I can't even get a crab licence or nothing for my longliner. I've been at crabs before with my father ... I tried to get a crab licence. I've tried for seven years. I lost $4700 on taxes last year because I had the longliner. The boat burned $19,000 on fuel. And I put insurance on the boat for $900 per month for six months ... Part-timers should have no salmon licences and pensioners shouldn't have 'em either ... The fishery has no future 'cause there are too many people involved with it, and you get schoolteachers there trying to make a quick dollar. I got the boat 'cause I thought things would brighten up. They give me the loan for the boat but not a licence to fish with her. (Respondent no. 048)

These comments were echoed, with greater resignation, by another longliner skipper:

To get a good living today, you've got to have a restricted licence. But the gill-netters don't have much of a chance. (Respondent no. 062)

Finally, a full-time fisherman in his sixties, who had given up the trapboat to pool his resources with his son in a longliner, provided this poignant vision of dashed hopes and the onset of despair:

My son has a longliner. Before I fished with him I was fishing with the Cold Storage and with Arctic Fisheries, and I was fishing on longliners with both of them ... When my son got a longliner, we all went out together. We thought we'ed do better that way. Now he got to sell it. The gill-netting is not paying for it. He got her on a loan and is having a hard job paying it back ... We're not using the longliner now. We uses a trapboat. (Respondent no. 035)

Under such circumstances, it is no wonder that those who continue to operate longliners are obliged to flood the fishing grounds with gill nets in a desperate attempt to recoup their own losses.

Many of the longliner operators just described had acquired their boats in the hope of using them for crab fishing, but found themselves caught in something of a catch-22. Without a longliner, they had no hope of getting a crab licence; owning the boat, however, was no guarantee of getting one. And, as we have seen, without a crab licence, at least in Bonavista, owning a longliner was a virtual recipe for bankruptcy. This does not constitute the unregulated competition of all against all depicted by the theorists. Rather, it is a competition that takes place in a vortex of regulations that are, in themselves, both the cause and the consequence of most of the competition.

'Crabmen got it made.' (Respondent no. 049)

When Bonavista's small-boat and longliner fishermen dream, they dream of owning a crab boat (that is, a vessel large enough to enable them to enter the crab fishery, together with the licence permitting them to do so). One of the fishermen quoted at the start of this chapter stated that there were 'three kinds of fishermen' in Bonavista, and there can be no doubt that, in everyone's mind, it is the 'crabmen' who have 'got it made':

The only thing is that a man with a crab licence is doing all right. There is room for more crab licences here if they land somewhere else to put out the crabs [i.e., the local plant is already operating at capacity]. (Respondent no. 035)

In Newfoundland fishing parlance, a 'highliner' is a fisherman who is highly successful and financially secure. In Bonavista, crab fishermen are described as highliners. Our data certainly support that characterization. The crab-boat skippers, and even some of the ordinary crab-boat crewmen, make considerably more money annually than the most skilful and lucky of the small-boat and longliner fishermen.

The financial arrangements that exist among the crews of privately owned crab boats are similar to those we have already encountered in relation to other types of fishing enterprises. The boat itself is owned by a skipper,[4] who, by virtue of the loan agreements into which he has entered, shoulders a considerable amount of the risk if he cannot make the required payments. As in the case of the small boats, a share of the income is therefore paid to 'the boat,' but in this case, because the costs of financing and maintaining the boat and its equipment are so high, the boat share is actually the largest. For each of the privately owned boats from Bonavista on which we obtained data, the boat share was 40 per cent of the income from the catch. That the crewmen also bear a portion of the risk is reflected in their receiving shares of the income from the catch as well. Thus, on each of the privately owned boats we studied, the remaining 60 per cent of earnings from the catch was divided equally among the crew members, including the skipper. Crews varied from five to six men, with the result that each person's share amounted to either 12 or 10 per cent.

It is of particular interest that the skipper's share was no larger than that of any of his crewmen. Unless he was able to make a profit from the boat share, his income would be no higher than any of theirs. In the words of one skipper,

The boat takes 40 per cent, and 60 per cent takes five of us – that's 12.5 per cent. I pays for everything see. When it ends up – the crew is just as good as me. (Respondent no. 061)

To be sure, some skippers have fewer debts and do better than others. Thus, one skipper reported receiving more than his 10 per cent share:

We share. Each crew member takes 10 per cent and I take 50 per cent [i.e., 40 per cent boat share and 10 per cent personal share]. For my personal living I take 10 per cent plus 5 per cent commission. That's 35 per cent operation. They tell me that's more than good. (Respondent no. 046).

Indeed, this particular skipper estimated his annual income from crab sales alone at well over $30,000 in the previous year, to which must be added his unemployment insurance and the additional income he received from everything from salmon fishing to sealing. Another skipper estimated that his total income from fishing was '$50,000 or somewhere around there, after I paid me expenses.' He was quick to add that, in order to travel farther offshore, he was about to purchase a new boat and would then take on considerable additional debt:

We get a new boat, and we won't be clearing that much. (Respondent no. 061)

In stark contrast, however, was a skipper whose expenses far exceeded his boat share of the income, with the result that his total income from fishing in the preceding year had been no more than $12,000. The crew members whom we interviewed had incomes ranging from $20,000 to $25,000.

The preceding discussion pertained to privately owned crab boats.

However, five of the crab licences allocated to Bonavista were held not by individuals but by the local crab-processing plant. In at least three cases, the boats being used in conjunction with these licences were also the property of the plant. The skipper of one such boat explained:

The [firm] has two more crab boats besides the one we use. One is about the same size as ours [i.e., 58 feet / 18 metres] and the other one is 52 to 54 feet [16 to 16.5 metres]. (Respondent no. 046)

Another company skipper described how he came to be hired by the processing company as follows:

The company needed a man and the fellas down there were giving trouble, and the plant manager phoned me [a skipper of a longliner without a crab licence] and he asked me to take it over. (Respondent no. 062)

The third company skipper operated from his own longliner but used the processing firm's crab licence:

I owns my boat, but I am using the company licence 'cause their boat is up for repairs. The ways things are going now, they are thinking of transferring the licences. They needs the boat 'cause they would have lost their licence somehow if they never had it. (Respondent no. 069)

The skipper of a privately owned crab boat was critical of this arrangement:

The company got five licences for five boats. They can take any man [for skipper] they want, if they wants to, and lay off the fella they got, and that fella got no crab licence ... They got one boat the [i.e., this] year chartered. That's still a licence belonging to the company. He [i.e., the skipper] don't have the licence. He's using the company licence while he's chartered. They can take it next year and give it to somebody else if they wants to. (Respondent no. 061)

The share arrangements on these company boats differed somewhat from those on privately owned crab boats, for the skippers were employees rather than owner-operators. In these cases, 35 per cent of the money went to the boat, 5 per cent was the 'skipper's commissions,' and the remaining 60 per cent was divided equally among the skipper and the crew members. However, lest this be seen as a particularly good deal for company skippers compared with private skippers, it should be pointed out that both the company skippers and the private operators were unanimous in their condemnation of the age of the company boats and the poor equipment on them. They stressed that such factors had a negative effect on earnings. In the words of two of the company skippers,

The only thing it [i.e., the company boat] is any good for is scrap. The only thing that's any good is the engine. I was off twelve miles and I had to fix the power take-off and I went to start the engine up again, and well, it wouldn't start. Good thing a fella was close by to tow me in. Good thing also I wasn't seventy miles offshore. [The boat] is twenty years old ... has no CB, no LOREN [i.e., a satellite directional-guidance system common on private crab boats and many longliners], no RDF [direction finder]. The radar's good for five miles although it's supposed to be good for twenty-four miles. (Respondent no. 070)

The company owns the boats and every year there got to be repairs – and the company doesn't start repairing them until they have been out fishing. This year [a private skipper] had 220,000 pounds of crab landed before we started. I say we lost 100,000 to 150,000 pounds of crab the [i.e., this] year ... The private boats here have better equipment than us ... I've been after them for three years for better equipment – promises and promises. (Respondent no. 062)

Another company skipper explained that, because of the poor condition of his boat, he was often limited to fishing unproductive nearshore waters, whereas the newer and better-equipped private boats were able to venture farther from shore:

Coming on now when the weather gets bad, we are almost forced to give it up – it's too hard to go out sixty to seventy miles ... The crab in the area we fish are not so thick [i.e., plentiful]. (Respondent no. 062)

One independent skipper went so far as to assert that the poor condition of the company boats made it difficult for the plant to find skippers willing to take them over:

The way it is with the boats that they got now, it is a job to get a skipper to go on them [because they are] eighteen to twenty years old. (Respondent no. 061)

However, this independent skipper's rather haughty attitude is not confirmed in fact; indeed, many a longliner operator would leap at the opportunity to skipper or even work on a company boat.

The fact that the company owns private crab licences gives rise to considerable resentment throughout the Bonavista fishing community. There would appear to be good reason for this. Not only does the practice seem to be in direct violation of the general policy and practice of species-fishery licensing on Canada's East Coast, but it means that the company holds at least three of potentially the most lucrative fishing licences in Newfoundland – and this at a time when independent longliner operators are unable to obtain such licences and are consequently slipping slowly into bankruptcy. Even some of the company skippers opposed the company's ownership and control of this critical asset:

One thing I don't agree with is the company having five to six licences. I feel only the skipper should have the licences. It would be better for all of the crew and the company. (Respondent no. 062)

In a similar vein, a skipper of a privately owned boat proclaimed:

There's no other plant as far as I know got crab licences. I think they should be made to give it up. (Respondent no. 060)

A crew member on one of the private boats, who would presumably dearly love to skipper his own boat if a licence were available, was firm in his opposition:

The company shouldn't own crab boats. Why should the company have all the licences and other fellas not have it? (Respondent no. 057)

Viewed more abstractly, a species licence can be considered akin to a means of production. As such, company ownership of that licence is tantamount to a loss of control by the fishermen of this very lucrative and symbolically important means of production.

The preceding analysis has indicated that there are significant differences among crab fishermen themselves. They stem not only from the different types of ownership of the means of production (privately owned boats versus company-owned boats), but also from the very different fishing conditions that apply on the company boats as opposed to the private boats.

Because the company boats were in a state of relative disrepair and had outmoded equipment, the company crab fishermen fished largely in inland waters near the entrance of Bonavista Bay. The following is a description by one of the company skippers of a typical voyage:

We leave Bonavista at four [A.M.] and we takes over the pots, and we leave again [i.e., to come home] around 12:30 to one o'clock. And we comes back and we get home here ... at six [P.M.]. If it's windy, it will probably take us to eight o'clock. (Respondent no. 070)

In contrast, the privately owned boats, with their better equipment, venture well beyond the local Bonavista Bay area. They search out the larger beds of deep-water crabs that lie some 80 to 120 kilometres offshore:

There's a lot of 'em [i.e., crabs] out eighty to ninety miles, but you need a big boat for that. We are off fishing seventy-five miles southeast – about fifty to sixty miles off from St John's. (Respondent no. 060)

Our fishing grounds are fifty miles off St John's, almost. (Respondent no. 046)

Not only do such voyages take these boats out of harbour for two to three days at a time, but they fish up to eighty kilometres farther out than the 'offshore' deep-sea draggers. The fishermen are quick to point out the irony in the fact that their relatively small, fifty-foot boats operate at greater distances from shore and often in rougher seas than do the much larger steel vessels:

We goes out seventy-five miles. About twenty-five miles past the draggers. I passes the area where I used to fish when I was fishing on the draggers. (Respondent no. 045)

These fishing practices are not the result of some love of danger or yen for hardship on the part of the skippers of the privately owned boats. Rather, they are the consequence of a declining crab stock in the Bonavista Bay area:

There's no crabs in the bay. There was lots there when we started. We started in the bay and you could bring in 12,000 pounds then if the plant could take them. (Respondent no. 060)

Bonavista Bay is just about finished for crabs. A few in the spring and she cuts right out [i.e., ends]. There's not many boats in the bay. And gill-nets take up just as much crab as the (crab) boats. In a fleet of gear [i.e., gill nets] there'll be a thousand pounds of crab per haul. They shouldn't be allowed to put gill nets on crab grounds. They'll only throw them away. There should be a quota or closed season [for crabs] on Bonavista Bay. When we started first, Bonavista Bay was good. Now there's nothing. (Respondent no. 061)

The crab in the area are not so thick [i.e., plentiful]. We got to go farther and farther out each year. (Respondent no. 062)

However, the crews who fish eighty kilometres offshore are concerned that those grounds are being overfished as well, and that within six or eight years they too will be depleted. This is the result, in part, of the intense competition among boats from Bonavista and those out of St John's and Port de Grave in Conception Bay, all fishing in this relatively small offshore area:

The inshore fishery will soon be a thing of the past because draggers and other boats are taking everything the crabs like to eat. Everything else are being depleted. Boats in St John's have been catching one million pounds a year. We haven't got that much, and the whole lot of crabs will be gone in six to seven years. (Respondent no. 056).

The crabs are getting scarce. We're only making a living. I think there's enough into the fishery now. Any more boats and none of us would [make a living]. (Respondent no. 061)

Faced with an almost defunct crab fishery in Bonavista Bay and what they perceive to be a declining stock of crab on their present offshore fishing grounds, Bonavista Bay crab fishermen are faced with two alternatives. One is to invest in even bigger boats that would allow them to fish farther offshore, where bountiful crab beds still exist:

There's lots of 'em out eighty to ninety miles, but you need a big boat for that. (Respondent no. 060)

A fifty-eight footer is not big enough for where I've got to go. I'm going to try for a larger boat. (Respondent no. 061)

However, that alternative is risky financially, given the excessive debt load that moving to a larger boat entails. That debt load could be particularly high if, as was reported by one respondent, Fishery Loans Board regulations favour gill-net longliners over crab boats:

We can't get a subsidy for crab boats as much as the fellow who is a gill-net fellow. You can get 15 per cent, but you can't get 35 per cent I don't think. (Respondent no. 061)

Furthermore, the alternative of buying a bigger boat may simply be impossible in view of provincial government regulations that require crab boats to be approved as such. Thus, a skipper who sells his crab boat in the hope of securing the down payment needed on a larger boat may run the risk of not being able to get that new boat accepted for the crab fishery:

My boat is licensed too. If I sells her, I may never get another licence. Before I sells mine I got to check with fish licences. Me and [another crab skipper] tried last year and we couldn't get a licence for a sixty-five-foot boat. That's no good for us. We wants more than a fifty-foot boat [i.e., his present vessel]. (Respondent no. 061)

Finally, even if a crab fisherman wants to risk selling his existing boat in the hope of getting a licence for a new one, it is unlikely that he will be able to find a buyer. After all, how many people are likely to be willing to run the risk of purchasing a fully equipped crab boat when they may not be able to get a crab species licence to use it?

The second alternative open to Bonavista Bay crab fishermen is to expand into other areas nearer shore where there has not previously been an extensive crab fishery. This alternative is of particular interest to those who operate the older company boats that cannot go out even as far as the present offshore crab grounds. They are especially interested in moving northeast along the shore into the adjoining Notre Dame Bay and White Bay. This interest has been sparked by reports from crab-boat skippers operating out of Valleyfield, on the west side of Bonavista Bay, who have occasionally fished in these areas:

The Vallyfield crowd moved north of Cape Freels ... They got plenty [of crab] there. They say there's lots of crab there and no one fishing it. (Respondent no. 060)

However, here again, Bonavista crab fishermen are blocked by government regulations. Just prior to our interviews with these crab skippers, provincial fisheries officials divided the northeast coast of Newfoundland into two crab-fishing zones – one south of Cape Freels (which is the point of land on the western side of Bonavista Bay that separates it from Notre Dame Bay) and the other north of Cape Freels. This regulation appears to have been introduced over the direct opposition of the government's own crab advisory board. One Bonavista skipper who was a member of that board stated:

It never passed the government crab advisory committee or anything. The north–south boundary – all the fishermen objected to it and all plant operators objected to it. (Respondent no. 046)

Indeed, Bonavista crab skippers did not know of this regulation until they received their annual licence with the notation that they were restricted to fishing south of the Cape Freels north–south boundary line.

This year it was printed on our licence. I was licenced for south of Cape Freels. (Respondent no. 046)

The imposition of this limit came as a considerable shock to at least one Bonavista crab-boat skipper who had already made plans to fish in the Notre Dame Bay area:

There's no crab here. Now we can't fish north of Cape Freels. We was going to move into Notre Dame Bay this year. Now what are we going to do? (Respondent no. 060)

Crab boats from other communities south of Cape Freels reportedly attempted to violate the ban, but were 'driven back' by federal fisheries patrol boats:

One fellow from Vallyfield and one from St John's, they were drove up out of it this year. I can't see if we got a crab licence why we can't go fish where the crab is at. (Respondent no. 061)

Other Bonavista crab skippers simply protested verbally this further restriction on their fishing operation:

I was at crab before they gave out licences ... It's bullshit, I can't see why we can't fish where we like. Yes, I've fished up there and fished in Lewisporte and trucked me crab to Bonavista. (Respondent no. 053)

Despite their relatively high incomes compared with other inshore fishermen's many of the crab-boat skippers we interviewed did not speak well of the future of their industry in the light of the new regulation:

While crabs are getting short, we can't go anywhere else. It is better for them to have a quota on it. (Respondent no. 056)

If there is plenty of crabs I say you won't be able to ship it anywhere, we were fishing at St John's last year. The crab isn't what you call plentiful, but we couldn't ship it at all because the plant's capacity was full. But this year the demand is high and the supply is low. (Respondent no. 057)

I don't agree with the Cape Freels boundary restriction. It doesn't affect us now, but in years from now it might. We got to go sixty to seventy miles out now. But if the Cape Freels [regulation] wasn't in [effect] you could fish there ... I feel this Cape Freels thing will affect us fellas down the road. Say I wanted to go to Fortune or Placentia bays. I can't go there because my licence is restricted. There has been some guys talking about it with the union to see if they could get the Cape Freels thing changed again. The fellas from Notre Dame Bay can come in here and fish capelin. So there is no difference when it come down to the final point. (Respondent no. 049)

Ironically, while other inshore fishermen in Bonavista may think that 'the crabmen got it made,' that was far from a majority opinion among the crab skippers we interviewed. Although their incomes may have been adequate for the moment, they saw little to indicate that their income would be similarly secure in the future. Certainly, not all of their problems can be blamed on government regulations. Nevertheless, from their point of view, those regulations do virtually nothing to facilitate their attempts to earn a living in the fishery or to ease their problems in that regard.

'That berth was me grandfather's, then me father's,
mine, and now me son's.' (Respondent no. 056)

As we have already seen in the case of Bonavista and other communities, the inshore fishery is a competitive enterprise in which men using different types of technologies in different kinds of boats compete with one another for a scarce commodity as well as for the income to be derived from the sale of it. The essence of the competition in this fishery, we have argued, is not the war of all against all envisaged by the proponents of 'state of nature' common-property perspectives. Rather, the essence of the competition resides in the fact that the fishery is fundamentally a 'hunt' conducted according to a set of federal and provincial regulations, on the one hand, and in the context of traditional 'community laws,' on the other. Indeed, we have argued that these traditional community regulations were instituted by the fishermen themselves to control the conflicts that would otherwise erupt. On this basis, we have implied that common property as conceived in the 'tragedy of the commons' model – that is, as a resource available for open exploitation by any individual – simply does not exist. Instead, community regulations have transformed virtually all common property in the inshore fishery into collective property. Earlier, we described this collective property as, essentially, community property. However, as we shall see in more detail in our analysis of property regulations in Bonavista, some community-based property relations derive their regulatory authority directly from the community, and control rests at the community level, while others are largely within the control of the extended family unit (with the sanction of community norms).

To this point, this chapter has differed somewhat from the last two. Our discussion of smaller communities in the earlier chapters focused first and foremost on the extent to which internal community regulation of the fishery was intermeshed or conflicted with the external regulation of it by the various federal and provincial fisheries agencies. In contrast, the very size and complexity of the Bonavista fishery has required us to focus initially on the 'three kinds of fishermen' in the community – the small-boat line and trap fishermen, the longliner gill-net fishermen, and the crab-boat fishermen. Nevertheless, much of our discussion of each of these three fisheries has, of necessity, emphasized the extent to which it is intertwined with and influenced by a nexus of regulations that serve to structure it, limit the activities of those who prosecute it, and affect their ultimate success within it. Indeed, it has often seemed that the fishermen themselves are as entangled in a net of regulations as the fish they hunt become entangled in the fishermen's own nets.

It is now time to turn our attention away from the external regulation of the Bonavista fishery and to focus more directly on the internal, community-based normative structure that has also been developed to regulate it. Each of the four communities that we considered earlier demonstrated some form of community-based fishery regulation governing the location of cod traps. To be sure, not all fishermen recognized these regulations as such. We noted, for example, that the fishermen in the small communities of Charleston and King's Cove responded unanimously that they did *not* have 'some way of controlling where nets and traps are set in the water.' On further inquiry, however, we found that such locations were treated, essentially, as family property more than community property, in that they were passed on from father to son. We argued that, whether it was recognized as such or not, this system of inheritance constituted a traditional, community-based regulatory system designed primarily to control the competition inherent in the hunt for fish. However, as was clear from our interviews, this traditional inheritance system was not without problems. Occasionally, it was challenged by one fisherman's putting his nets closer to those of another than community custom allowed.

The two larger fishing communities, Grates Cove and Fermeuse, also had a community-based way of regulating fishing rights and the location of nets. In both, however, the practice was much more formalized, involving an annual community lottery 'draw' for fishing berths and a formal set of regulations specifying when gill nets had to be removed from trap berths. In both communities, the draw was familiar to all community members, with the result that all our interviewees there answered in the affirmative when asked whether their community had a way to 'control where nets and traps are set in the water.' Moreover, in both Grates Cove and Fermeuse, there was little evidence of the conflict over fishing berths that was apparent in the smaller communities. In Fermeuse, however, there was open conflict with fishermen from a nearby community over fishing grounds used in common by both communities. This suggested that intercommunity rivalry still remained a major obstacle to the peaceful local control of community common property.

Unlike the fishermen in the four communities already discussed, who were unanimous in their responses about the presence or absence of community regulations pertaining to the location of nets and traps, Bonavista fishermen were divided in their answers to the question. Twenty-five (58.1 per cent) stated that such community regulations existed, thirteen (30.2 per cent) claimed they did not, and five (11.6 per cent) said they didn't know whether their community had such regulations or not. This 'confusion' would appear to result from the usual nature of community regulation in Bonavista. Despite its larger size, Bonavista does not have a draw system for cod-trap berths. Consequently, some of the newer or less-experienced fishermen seemed to think there were no regulations at all:

There's lots of places and anybody will help you out. You just put out your marker. (Respondent no. 050)

Who is ever to the place first puts their traps into the water. (Respondent no. 029)

Just put 'em where you think the fish are to. (Respondent no. 028)

Other, more-experienced or better-informed, fishermen were adamant that cod-trap berths were allocated by inheritance.

That berth was me grandfather's, then me father's, mine, and now me son's. (Respondent no. 056)

I fishes the same berth as always. The same berth father used to fish. (Respondent no. 059)

Some people get their trap berths and have 'em every year after year. Nobody tries to take anybody's berth. (Respondent no. 058)

Of the four communities we studied, it was Charleston and King's Cove that had retained the traditional practice of allocating trap berths by inheritance. In contrast, both Grates Cove and Fermeuse had instituted an annual lottery system, or 'draw.' Although Grates Cove and Fermeuse were not much larger than Charleston and King's Cove, they had a fishery that was clearly the focus of economic activity in the community. In particular, there was substantial differentiation in the types of gear used to catch fish and the types of boats employed to harvest them in the two larger communities. One conclusion that might therefore have been drawn from our analysis of the four communities is that, as the fishery in a community grows and becomes differentiated, there is an increased likelihood that it will move to a lottery basis for assigning fishing locations. Although that proposition may still have some validity, our findings for Bonavista provide the 'deviant case' establishing that it is certainly not universally applicable. Bonavista probably has the largest and most highly differentiated inshore fishery to be found anywhere, yet it retains the custom of berth allocation by inheritance, which sees fishing grounds as family property rather than community property.

It is difficult to characterize succinctly the exact nature of these property relations. Certainly, as noted, there is no draw for trap berths in Bonavista, and a majority of the fishermen there recognize a right of families to pass berths down from generation to generation:

We only started two years ago, so we got berths that other groups don't want. You can't get a berth licensed now, there is a freeze on it. (Respondent no. 029)

At the same time, there also appears to be a shared understanding that, if a family does not occupy its berth early in the season, that berth can be appropriated by other trap crews:

Oh yes, [we] got pretty much the same place every year. But once in a while you couldn't get there in time. And if a fellow got your berth, there's nothing you can do about it, you know. (Respondent no. 071)

We keep it because we're the first there. We don't take other fellows' berths here. (Respondent no. 056)

If you can take it from someone else you can have it. They never draws in this area. (Respondent no. 030)

Thus, perhaps because there is no mechanism to enforce them, inheritance rights are tempered to some degree by a first-come–first-served principle. Unless one is quick to occupy one's inherited and traditional fishing berth, it will be taken by others, who will claim vociferously that they understood it to have reverted to common property. In short, under a traditional regulatory system such as is found in Bonavista, the rights to property must always be jealously guarded, for they can easily be lost. Is it any wonder, then, that our respondents were divided in their answers to the question about allocation of fishing berths?

It is clearly important for families to lay claim to their 'traditional property' as early in the fishing season as possible:

You keep 'em the best ways you can. The fastest man gets it. (Respondent no. 034)

Indeed, it is evident from our interviewees' comments that those who hold traditional berths must mark them in some way even before ice and sea conditions make it safe to do so. In order not to risk their expensive traps, which, if damaged or destroyed, would effectively ruin their fishing season, Bonavista fishermen have developed an elaborate set of techniques for laying claim to their berths.

The simplest way is to put out a set of moorings in the very early spring, indicating the general area in which one's trap will be located:

In the spring when they are putting them out – they puts down their buoys to set a place for their traps. (Respondents no. 026)

The way it is here in Bonavista is that the people with the good berths go out early in the spring and put out a mark. (Respondent no. 064)

But fellows go out in February – some fellas – to set up buoys for cod-trap rights. (Respondent no. 028)

It's not as bad today by half as twenty years ago. Then, you'd have to have your net out in April month. Now, all you can do is put out a marker. (Respondent no. 040)

Most fishermen use a set of moorings, to mark off the general area within which their traps will ultimately be located, but an occasional fisherman will take a chance, staking his claim with only one or two moorings and anchors:

This year I held berth just by two moorings. (Respondent no. 056)

Some of the fishermen who are lucky enough to have salmon licence put their salmon nets out in the locations of their prime cod-trap berths quite early in the season, in order not to lose those berths:

Fellows go out and put herring nets and salmon nets to hold trap berths. (Respondent no. 050)

This might not be the most effective use of a salmon licence, as salmon may not be plentiful in the specific locations that cod frequent. However, the total income from a salmon catch would be relatively small compared with the combined income and unemployment insurance that would be lost if one's prime cod-trap berth was occupied by someone else. Those who do not have a salmon licence might use old cod gill nets or other similar gear to accomplish the same purpose:

They has their own berths. Fellas carries down old gill nets or moorings to save their berth (Respondent no. 045)

One exceptionally resourceful crew has gone to the effort of having a 'fake' trap made in order to hold a prime berth until the trap season begins:

We got a trap that's made just on purpose for that. It's a small one we made just for berths and that – only seven fathom deep. The others are at least ten. We got three more bigger ones. If we puts the small one out early and she's damaged, then we haven't lost anything. (Respondent no. 040)

The apparent necessity of such forms of subterfuge invites closer consideration of the nature of the property rights embodied in such practices. We referred to inheritance rights to trap berths earlier as family rights, largely because the control of the property rests neither with the community *per se* nor the individual. The community, on the one hand, has no regulatory structure to govern such rights, as it does in the case of a lottery or 'draw' system. The most that the community can do is positively sanction traditional arrangements or frown upon those who do not comply with them. However, such reactions involve no penalties beyond those that might accompany negative public opinion. On the other hand, it is also clear that such rights are not individual, private-property rights. If they were, it would be generally accepted that an individual fisherman had the right to either rent or sell a fishing berth to another fisherman. It is quite obvious, however, that such a possibility has never even been considered by the Bonavista fishermen. As was implied in some of the comments by our respondents cited earlier, and as is emphasized in the comment that follows, a berth becomes open and available to others when there is no son in the family interested in occupying it:

Most berths are handed down [i.e., from father to son]. But when a man dies and his son doesn't want to take it, then someone else takes it. (Respondent no. 059)

The rights described here are quite different from the 'perimeter-defended rights' in the Maine lobster fishery described by Acheson (1975, 1987) and discussed here in chapter 4 (see pages 79–81). The Maine lobster rights were definitely private-property rights rather than common-property rights, because local residents accepted the right of individuals to rent and sell their 'fishing rights.' In contrast, the family-property rights in Bonavista (as in Charleston and King's Cove) retain their common-property character. Furthermore, this common-property character is best thought of as a 'modified family common property.' It is *family common property* as opposed to either *open common property* (because it is not the open, first-come–first-served scenario depicted by the 'tragedy of the commons' theorists) or *community common property* (because the community does not have a formalized system for allocating berths, such as a lottery draw, and because the community *per se* has no mechanisms in place for policing and regulating such practices). The nature of these rights as family common property, however, is clearly limited (or 'modified') in the sense that the right to control the property can be lost if the property is not occupied by a family member at the very start of the fishing season. When that occurs, the property appears to revert briefly to an open-common-property condition before becoming the family common property of another family. The Bonavista example thus demonstrates that situations involving open-access natural resources, such as the fishery, can and do become moulded into a wide range of common-property arrangements and that those arrangements can be transformed from one type to another under the aegis of a set of community norms of varying formality.

'There's too much self. They wants it all.'
(Respondent no. 037)

The preceding conceptual analysis of the nature and types of common property rights does not adequately address one of the fundamental questions – namely, why a diversified fishing community such as Bonavista, with all the complexity that clearly exists in its social relations, should retain a traditional form of property relations based on the principle of inheritance that appears more in keeping with less-differentiated fishing societies. When questioned about why Bonavista had not switched to a draw system such as those found in other Newfoundland fishing communities, some of the local fishermen argued that it was because Bonavista's headland location resulted in their having to fish in two different bays – Bonavista Bay and Trinity Bay – in which the depth of water differed so greatly that a draw was not feasible:

Can't do it here, It's the depth of the water, see. You get eight feet of water one place and twelve feet another. If I got a twelve-foot trap and draws an eight-foot berth, how do I use that? (Respondent no. 056)

But that rationale does not (pardon the expression) hold water, in view of the data that we have available from Grates Cove. Like Bonavista, Grates Cove is situated on a headland and its fishermen find themselves fishing in waters of quite widely varying depths, yet the lottery-draw system of berth allocation does not pose an insurmountable problem for them. Draws for berths in Grates Cove are held sufficiently early in the winter season that fishermen are able to shape their traps to fit the depths of their berths.

In our judgment, Bonavista's maintenance of the traditional, inheritance-based approach to the allocation of berths is attributable to two factors. One is simply the intensity of the competition that exists among the fishermen of Bonavista. Indeed, the competition seems to be so great that it allows for very little cooperation – too little even to change the existing system. Some evidence of the intensity of this conflict is to be found in the already-cited comments of the fishermen themselves. But perhaps the most poignant statement of that conflict is contained in the following quotation:

We draw for nothing here. The only place around the island. There's too much self. They wants it all. There's fellow with three or four traps out and no one else will get it. They goes out in March and puts out mooring or gill nets to hold them back. That can't hold a berth. There's no way that can hold a berth. If I had a trap I'd have a berth. There's no man would stop me with three or four traps. They should be 'drawed for' here like they does in Catalina ... Some fellas never gets good berths. The traps all in good berths are down in the tickle between Green Island and the Cape. Some fellows never gets nothing where they got 'em. (Respondent no. 037)

The second, perhaps even more important, factor contributing to the persistence of traditional practice in Bonavista is the intense competition between its fishermen and fishermen from other, nearby communities, particularly in Trinity Bay, for access to the same fishing grounds. The following statements capture some of the intensity of the intercommunity conflict:

If we were not allowed to fish in Trinity Bay for cod, I'd starve. (Respondent no. 056)

We was [fishing] way up between Ellison and Catalina [two Trinity Bay fishing communities]. There's nothing they can do about it either. Even though they can draw up there. But it's not that way using traps here. (Respondent no. 034)

It is our argument that the absence of a formalized, community-based system of regulation for the trap fishery in Bonavista stems from the fact that the fishery in Bonavista is not really a 'community fishery' at all. Rather, it is the extensive exploitation of a broad fishing territory that overlaps the fishing territories of many other fishing communities, both nearby and distant. In such circumstances, developing local fishing regulations for one's own community might mean having to recognize the legitimacy of other communities' territorial rights over fishing grounds currently exploited regularly by one's own community's fishermen. As the preceding quotations reveal, communities in Trinity Bay that have lottery draws for fishing berths cannot enforce the resulting property 'claims' against the Bonavista fishermen, particularly when the latter assert their traditional right, inherited from their forefathers, to fish these other communities' fishing grounds. However, if Bonavista switched to a lottery-draw system, it might have to forgo such traditional claims, and it would probably also have to limit its fishing territory. Certainly, it would be difficult to publicly legitimate a process that would allow Bonavista fishermen to draw for fishing berths that were in no definable way a part of their local fishing grounds but that were obviously a part of some other community's. In sum, a traditional berth-allocation practice persists in Bonavista partly because the community is too divided and fractious to agree on a mechanism to change it, but more likely because any formalized attempt to regulate the practice would be likely to result in a diminished fishing area. Considered from the opposite perspective, this somewhat anachronistic practice remains in place because the community as a whole benefits more from it than it would from any alternative system of local regulation. In short, it is functional.

Those who hold traditionally allocated, or inherited, trap berths tend to strongly favour the practice, even when, occasionally, they lose out to others in the race to occupy their usual berth:

We've had berths taken from us several times and put ours somewhere else and done just as good. We got lots of berths for those trapping now. I know of four good berths this year with no trap there, and I'm supposed to have the best trap berths in Bonavista this year. After forty years you knows the bottom and where to put your traps. (Respondent no. 040)

However, while the traditional practices may be functional for those who have inherited berths, they are not functional for everyone. Notably, they discriminate against those who wish to enter the trap fishery but who lack the 'family legitimated' rights of access to the best berths. As might be expected, such people are highly critical of the existing system:

I knows one thing I like to see done is to draw for trap berths so other fellas can get a chance. (Respondent no. 032)

They wouldn't draw at all, but drawing would be the right thing. (Respondent no. 067)

The fellas with the trap berths have had 'em for years. They should have a trap-berth draw. I can't get a decent place to put [my trap], and there are some fellas here with three trap berths, and they have 'em every year. They are the ones crying about us putting our gill nets even only a half-mile from their trap. And they still complain that we are cutting off the fish from them. (Respondent no. 039)

Oh yeah! People got their own berths and keep 'em for years. People got berths here since they were born. It's not right. (Respondent no. 068)

The best part of the berths are used up 'cause fellas save 'em from year to year ... You have to take others' berths if you wants a good one. (Respondent no. 045)

Other fishermen without traps point to the obvious potential of the traditional practice to intensify community conflict:

You don't use traps [i.e., enter the trap fishery] unless you want to take a chance and put 'em where you want to. But then you are making enemies. (Respondent no. 039)

Indeed, the inclination to avoid conflict that is evident in the preceding comment may, in itself, have, led to a decline in the trap fishery:

There used to be rows and bad friends, but not so bad now. There's less cod traps. Then they used to put out markers with old gear and somebody else get close to 'em and then have a row. I think they should draw for berths and if they wanted a change, you could. (Respondent no. 048)

Nevertheless, the potential for conflict over the possession of trap berths remains ever-present and is a far cry from the 'ideology of cooperation' and its accompanying rhetoric that one must usually penetrate before one can understand the actual workings of fishing communities. The willingness of some fishermen in Bonavista to criticize the traditional trap-berth allocation system openly again indicates that the 'reality of conflict' lying at the heart of most fishing communities is much closer to the surface there.

'You could have a gill net in a trap berth
then, but not now.'
(Respondent no. 027)

It would be misleading to leave the reader with the impression that there is no basis of community regulation in the Bonavista fishery. Although there may not be much community cooperation in the allocation of berths for the traditional trap fishery, there has been some, relatively recent, community effort to regulate the longliner gill-net fishery. The new community regulations were summarized by two fishermen as follows:

They had an agreement in 1971 on cod gill nets, whereby there is only a certain area you can set in, and a certain time – from July 1 to August 10. But there are areas outside of that you can fish any time. Those areas are too far out for small boats, and there is no trap berths set. (Respondent no. 028)

Well, eh, gill nets have a certain time to go in and a certain time to go out. July 1st to August 10th inshore, and salmon nets May 20th to end of the year. (Respondent no. 052)

In other words, shortly after the introduction of the longliner–gill-net fishery into Bonavista in the early 1970's, the fishermen of the community collectively agreed to place restrictions on the locations where fleets of gill nets could be set. The implementation of such measures was an obvious attempt by trap fishermen to control the encroachment of the longliner–gill-net fishery into the fishing grounds from which the trap fishermen drew their livelihood:

One year, about six or seven years ago, we had our nets out, a gill net in a trap berth. You could have a gill net in a trap berth then, but not now ... In order to have a cod-trap berth now, you got to have a cod trap. (Respondent no. 27)

This community-wide agreement to establish regulations governing the location of gill nets should not, however, be seen as a harbinger of increased cooperation among the fishermen of Bonavista. It is better seen as an agreement brought about by the efforts of trap fishermen to regulate the overcrowding of their traditional trap-fishing grounds, which had occurred as a result of the introduction of longliners and gill nets. Indeed, it might best be viewed as a deliberate effort by the trap fishermen to ensure that their livelihood would not be completely threatened by those who were adopting the government-encouraged move to longliners and gill nets. The instigation of this minimal level of formalized community regulation, from this perspective, is not an act of community cooperation, but an attempt by one group of fishermen to divide up the fishing territory around the community. Such regulations only hamper further the longliner owners' ability to make an adequate living.

'A lot of regulations don't make it no better.'
(Respondent no. 036)

The picture of the Bonavista fishery that emerges shows the community as having close to the minimum level of local cooperative regulation needed to prosecute a local fishery. We have suggested that this low level of cooperation has endured primarily because it is, at least from the perspective of Bonavista, a functional solution to the fact that the fishing territory claimed by Bonavista fishermen overlaps that of numerous other communities. It is also consistent with the general orientation of the community fishery in Bonavista, which, by and large, eschews the cooperative ideology that is found in many other communities.

However, as we have seen elsewhere, community regulation of the local fishery must take place within the context of the web of external regulations that government and other agencies have created. Of particular relevance in this regard are the roles of government and of the fishermen's own union. Thus, before concluding our discussion of the regulation of the Bonavista fishery, we shall examine how these external forces are regarded by its fishermen.

Although none of the communities we have examined has warmly embraced the government's regulation of its fishery, there is often a grudging recognition that some level of external control may be needed. Such sentiments are usually expressed in terms that echo the rationale of the 'tragedy of the commons' position, namely, that external regulation is needed to prevent the overexploitation of the resource. To be fair, similar sentiments were expressed by one Bonavista fisherman:

You've got to have control o' it. If there are too many fishing the one place, there will be none for anyone. (Respondent no. 062)

However, the overwhelming majority of Bonavista fishermen were clearly opposed to government regulation of their fishery. An indication of this negative appraisal has been evident in the various statements from trap-boat fishermen cited earlier that criticize government's encouragement of the longliner–gill-net fishery. We have similarly seen evidence of longliner fishermen's negative views of the difficulty they face in obtaining protected-species licences. Crab fishermen have been just as censorious of governmental regulations that prohibit them from fishing north of Cape Freels. Thus, it would appear that, if in nothing else, all three types of Bonavista fishermen are united in their condemnation of the way government regulations affect their access to the fishery.

Bonavista fishermen also echoed fishermen of the other communities we examined in their criticism of the other government policies. For example, there was general condemnation of the distinction between full-time and part-time licences:

No one should have to have a licence. It's only money for the government. If a man devotes his lifetime to the water, he shouldn't need a licence at all, or he shouldn't have any trouble getting one. No matter what kind of licence it is. (Respondent no. 070)

But with regard to the fishing licence, I don't know. They got so many fucking rules and regulations it is not funny. (Respondent no. 057)

Similarly, there was vehement opposition to 'moonlighters,' who, more often than not, were identified as schoolteachers:

There should be no part-timers at all. Either you are full time, or you are not. People around here now, schoolteachers, got licences, and they are not fishing for a living. People around here get jobs besides fishing. I don't call that a fisherman. There should be no part-time. (Respondent no. 058)

Also reminiscent of other communities were the many complaints we heard about the government policies pertaining to the allocation of species licences:

Part-timers should have no salmon licences and pensioners shouldn't have 'em either. (Respondent no. 058)

If a man is retired from fishing, he shouldn't have no salmon licence. (Respondent no. 056)

However, in other communities, such complaints were usually directed at specific policies of fisheries regulation. The somewhat disturbing aspect that set Bonavista apart in this regard was the blanket condemnation of all aspects of government involvement that we heard from a number of fishermen:

I don't see much sense to none of what the government does this day ... A lot of regulations don't make it no better. (Respondent no. 037)

With regards to the fishery, everything wants to be done and nothing is going to be done. The government will do what they wants [to do] anyway. (Respondent no. 025)

The level of alienation was so pronounced that some of our respondents believed the government was deliberately trying to destroy the inshore fishery in the province:

They [the government] is trying to kill the inshore fishery ... They'd sooner me be on a dragger than paying me twenty-eight cents a pound. (Respondent no. 056)

They're trying to shut down the inshore fishery. The inshore fishery puts up the best fish too. It's a hard proposition too. Fifty per cent of Newfoundland is in the inshore fishery. (Respondent no. 064)

Judging by such statements, it seems clear that the level of disaffection with regard to government involvement in the fishery was generally higher in Bonavista than in the other communities we studied.

In contrast, Bonavista fishermen's attitudes towards the role of their own union were more mixed. Some fishermen steadfastly maintained that they were union members only because the dues were automatically deducted from their income by the fish plant:

I joined it cause they wouldn't take n'ar squid from you. The way it is, they puts your name down and if your name's not there they wouldn't take n'ar squid. (Respondent no. 038)

No other choice. They automatically takes the dues out as long as you takes fish in. [i.e., sell fish to the plant]. (Respondent no. 036)

There were others, however, who listed in detail the benefits they derived from membership in the union:

They have [done some good] in certain insurances: wages-wise, price-wise, more or less on welfare benefits, medical benefits, and life insurance through the union. (Respondent no. 039)

A majority were quite willing to credit the union with securing higher prices for their fish:

I joined the union for a better treatment. Those in the fishery [i.e. merchants] could give you any price they liked. The union looked out for a better price. We wouldn't get the price we are getting now without the union. (Respondent no. 027)

You wouldn't be getting the price you are getting if you didn't get organized into the union. (Respondent no. 030)

This sentiment was shared even by the crab skippers and crews:

It has helped 'em price-ways. I don't think you'd have anything without the union. We wouldn't get twenty-eight cents per pound for crab. (Respondent no. 056)

In the light of this level of satisfaction, the quotation at the beginning of this chapter that concludes, 'the union is no good in Bonavista – it's got one fella penalizing the other,' may seem unrepresentative of Bonavista fishermen's views. But there is also considerable evidence in our data to indicate that Bonavista fishermen had not forgiven the union for involving them in a general fishermen's strike some years before – a strike that was ultimately settled for little benefit to the inshore fishermen:

I don't think it has helped too much. A couple of years ago we went on strike and the strike made us worse. You can't get back what you loses. (Respondent no. 052)

The other years we went on strike for twenty-three cents a pound. We went back for twenty-seven cents and ended up getting twenty-five cents. They cut the price to twenty-three cents per pound, so that is why we went out. Then they locked us out. (Respondent no. 062)

Some people says it is all right, but I can't see where they helped me out one bit. Because the other year they pleaded for us to go on strike and they got so many to vote for it to go on strike. And the man who didn't want to go on strike had to go on strike too. And when it was all over, we put in [i.e., were on strike] five weeks and we were back where we started, and we weren't as good as when we knocked off [i.e., went on strike]. (Respondent no. 035)

I think the union has got it all tangled up. The union wants us to strike for pay. The union wants more money and something else will go up. I think the union is wrong cause when we was on strike two years ago, we lost more than we gained. (Respondent no. 047)

There was also some evidence that the union was perceived as being of greater benefit to some types of fishermen than others:

They [the union] has helped the trawlerman more than anyone else. On a good trawler they average $20,000 per man. (Respondent no. 062)

I don't think it helped the crab fishermen that much. After the strike we lost two cents a pound. Yet we shouldn't have been on strike. It had nothing to do with us. But we couldn't cross picket lines. (Respondent no. 053)

And, finally, some fishermen voiced the suspicion that the union leadership was benefiting too much at the expense of the fishermen:

The union is only making a big haul for itself. We're getting nothing out of it. (Respondent no. 057)

A union can get a few cents for you, and they can starve you too. One fella, he fished all his life and he got a part-time fishing licence like hundreds in Bonavista. He's in the union, and they penalize him for not taking his fish [when there is a glut]. But a lot of men in the union committee has got boys with part-time licences and they [the fish plant] takes their sons' fish. But this other fella's a skipper, and they won't take his fish. (Respondent no. 070)

'When Bonavista goes out as a fishing settlement, there will be a lot of places that goes before it.' *(Respondent no. 067)*

In the other communities we studied, the majority of those we interviewed claimed to like their community and the quality of life it provided. Although our Bonavista respondents were not unanimous in their positive evaluation of their community, as our Grates Cove and Fermeuse respondents were, they nevertheless exhibited a high level of satisfaction (see Table 7.2). Forty-nine out of 50 respondents said they liked living in Bonavista:

There's no better place. (Respondent no. 057)

Similarly, 45 out of 50 thought it was a good place to raise children, and 46 out of 50 considered the people generally hard-working. There was also a relatively high level of agreement that the people in Bonavista 'cared about the place':

You can tell that by looking at the houses. (Respondent no. 067)

However, there was far greater uncertainty than in other communities about the quality of community leadership. Fewer than half the respondents gave the leaders a positive rating.

This negative attitude also extended to the level and quality of community services. Whereas Fermeuse and Grates Cove residents were satisfied with the level of services in their communities, Bonavista residents were generally critical of theirs, despite Bonavista's larger size. Considerable dissatisfaction was expressed about medical services; it

TABLE 7.2
Bonavista respondents' views of their community and its future

	Yes	No	Uncertain
Evaluation of community			
Like living here	49	1	0
Good place to raise children	45	1	4
People here generally hard-working	46	0	3
Leaders are good, capable people	22	10	15
People here care about the place	42	0	6
Evaluation of community services			
Good schools and teachers	42	1	7
Good medical care	29	13	8
Enough to do in spare time	22	18	5
People here do without a lot	19	17	12
Evaluation of employment opportunities			
Good place to find work	4	44	1
Fishery has a future here	21	12	15
Fish plant here has a future	20	10	19
Inshore fishery good for young people	15	23	10
Evaluation of community's future			
Children should settle here	30	3	6
Community has a good future	26	7	16

Note: There were 50 respondents. Where responses for any given item do not total 50, the balance represents abstentions.

centred largely on the distance by ambulance to St John's for major surgical care:

We got nothing here. We got a hospital to go to. Fellow went in for tonsils and they gave him a pill for heart. We got to go to St John's for everything. (Respondent no. 055)

I had to make too many trips to hospital [in St John's]. Three or four trips last year. (Respondent no. 064)

Others complained that, for even the most minor purchases, they often had to travel roughly a hundred kilometres, to Clarenville:

We got to go to Clarenville to buy a line and jigger. (Respondent no. 055)

Finally, we heard the usual complaint that the community lacked the recreational facilities that young people wanted:

Not much future for young people. There's not much for them here. (Respondent no. 071)

Especially young people. Nothing for young people. (Respondent no. 072)

   Furthermore, as was the case in the other fishing communities we studied, the residents of Bonavista were close to unanimous in their negative evaluation of employment prospects in their community. Forty-four out of the fifty declared Bonavista, in particular, and the fishery, in general, most definitely *not* to be 'a good place to find work':

I wouldn't advise a fellow to go at it. [i.e., the fishery] if he had a good education. Not the inshore racket anyway. (Respondent no. 068)

Everything is getting complicated, and it's hard to make a living. What's $10,000 [i.e., his last year's income]. Not much for the hours you put in. (Respondent no. 061)

Fewer than half of those we interviewed felt that either the fishery or the community fish plant had a future:

For a young fellow to go into the fishery, you need secure markets. But the market's price is something else. We haven't had a price increased in three years and that is going to hurt, with everything else going up. (Respondent no. 067)

Concern was expressed about the new owners and managers of the fish plant, in particular; they were virtually unanimously condemned for their management practices:

I don't like the way the plant is treating fishermen. Why should we suffer for the fish plant. They holds over fish till it's left over three or four days. By then its maggoty. It is almost walking in the curing room. (Respondent no. 055)

The [previous owners] were good managers. They were smart people, and anybody in Bonavista will tell you the same thing. (Respondent no. 059)

I doubt it with the [current owners]. They're pretty cheap. Take the fish plant out of Bonavista and it's a ghost town. (Respondent no. 070)

Some of the criticism clearly arose in response to the above-noted change in management, and in management style. The former owner was a local man who knew all the residents, was always available, and provided loans for nets at the start of the year:

When [the former owner] was here, you call him anytime of the night and he got help for you. It's not like that now. One time ... you could come into the plant at seven or eight o'clock, and if you got trouble with your boat, it would be fixed. Now if you don't get in by five o'clock, you miss a day fishing 'cause there is no one there to look after you. (Respondent no. 072)

In contrast, the new owners had not only eliminated many of the traditional practices, but had 'had the audacity' to bring in such innovations as requiring that fishermen take crushed ice aboard their boats to keep down the temperature of the fish they caught until it could be brought back for processing.

Such changes seemed only to increase the general pessimism of most Bonavista fishermen about the future of their community and its fishery. Although 60 per cent felt that their children should settle in the community, only slightly more than 50 per cent felt that their community had a good future. Many of those who did, clung to the belief that the community, along with its fishery, had a future simply because it was too big to be allowed to collapse:

When Bonavista goes out as a fishing settlement, there will be a lot of places that goes before it. We goes out on the continental shelf. With good management, this plant could stay on for a long time. (Respondent no. 067)

When the fishery is gone, it is all gone. (Respondent no. 068)

It's not got much future. But I guess the government will keep it going. (Respondent no. 072)

Conclusion

In each of the five community studies presented in this book we have tried to explore the same set of themes. In doing so, we have found ourselves following the same threads of argument. However, those threads have been woven into very different patterns in the five different communities.

Bonavista has certainly been no exception. Although our dominant concern has continued to be the interrelationship of community and state regulatory practices, we have found ourselves focusing on different aspects of fishing regulation in Bonavista than, for example, in Grates Cove and Fermeuse. In examining the latter two communities, we focused on the relationships between full- and part-time fishermen, between fishermen who hold species licences and those who do not, between fishermen who use different modes and different technologies of production, and between the control of common property and the unemployment insurance system, which provides some modicum of income security. In contrast, we mentioned the first two of these relationships only in passing in our discussion of Bonavista, and we did not consider the fourth at all. To be sure, this is the result, in part, of a deliberate effort to avoid redundancy. But it is also a reflection of the fact that the dominant issues facing fishermen and the fishery in Bonavista are quite different from those confronting the communities we considered earlier.

Two issues have dominated our consideration of fishing-community life in Bonavista. First, we have explored the relationships among three different types of inshore fisheries – the small-boat handline and trap fishery, the longliner–gill-net fishery, and the crab fishery. More than the other communities we studied, Bonavista demonstrated how these three fisheries tend to pit modes and technologies of production against one another and how the regulatory process that applies to one mode and technology inevitably affects the life chances of fishermen using other types. The second issue involves the factors contributing to Bonavista's apparent lack of interest in moving from a traditional form of trap-berth allocation to the more democratic, lottery-based system that was found in other larger and more-complex fishing communities. It is our conclusion that traditional allocation patterns remain in Bonavista because any deviation from them might require Bonavista fishermen to limit the extent to which they can counter other communities' competing claims with their own.

Finally, we should not conclude this chapter without some formal recognition of the fact that the fishermen of Bonavista, the largest inshore fishing community in Newfoundland and, perhaps, in the world, have little hope for the future of their fishery or their community. Any hope they do have seems to reside in the belief that Bonavista is 'too big' for the government to allow it to collapse. Although pessimism about the future of the fishery was certainly evident in the other communities we studied as well, it is particularly depressing to find it so firmly ensconced in what should be the pinnacle, the shining example, of the inshore-fishing industry. This, perhaps more than anything else, is clear evidence of the extent to which those who regulate the fishery have failed those who make their living from it.

8 Guiding Metaphors

Some Concluding Thoughts on
Future Fisheries Regulation

Guiding metaphors are powerful limits and directives for the resulting theories. They highlight that which conforms, and they suppress 'irrelevant aspects'; thus they can lead to a kind of 'tunnel vision.'

Scott Greer,
The Logic of Social Inquiry (1969, 148)

The universe would appear to be something like a cheese; it can be sliced in an infinite number of ways – and when one has chosen his own pattern of slicing, he finds that other men's cuts fall at the wrong place.

Kenneth Burke,
Permanence and Change (1954, 103)

At the beginning of chapter 3 we quoted the words of Archibald MacLeish: 'A world ends when its metaphor has died.' We implied there that fisheries policy, as indeed all policy, is guided not so much by facts or reality as by a representation of reality. In time, this representation comes to be accepted as reality itself. Put somewhat differently, under the influence of such metaphoric representations, facts take on distinctive meanings. However, lest metaphors be perceived as prisms that unnecessarily distort our understanding of reality, it should also be noted that facts, in and of themselves, indeed have no meaning. Facts take on meaning only in terms of the metaphoric perspectives according to which they are interpreted.

Within the sciences and social sciences there are many well-known metaphors. For example, electricity is metaphorically represented as 'flowing' like water, and diplomatic and battlefield behaviours have come to be understood in terms of 'game' strategies. Similarly, metaphors in the social sciences represent systems as being 'in equilibrium' and human behaviour as 'drama' and 'theatre.' In this book we have been particularly interested in the power of the metaphor that represents certain natural resources as being subject to the 'tragedy of the commons.'

The force of this metaphor is so great that it has been used as the basis of many of our most enduring theoretical justifications for the formation of political society. Hobbes and Locke postulated their theories about the rise of the state on the need for a civil authority to protect the right to property from the possessive individualism and exploitation that would otherwise prevail. Indeed, it is the right to property, in the guise of class interests, that forms the basis of much Western political theory, from Adam Smith to Karl Marx. In short, property relations, like class relations, are at the very foundation of civil society.

It is in this context that the significance of the present study for our understanding of the nature of fishing and fishing rights in the Canadian East Coast fishery can best be assessed. Perhaps its most important contribution lies in its demonstration of how the 'tragedy of commons' metaphor came to dominate Canadian policy makers' perspectives on property relations in the East Coast fishery. The regulation of the fishery came to be understood not so much as a biological problem relating to conservation and natural selection, but as a social problem relating to human greed and exploitation. Although, in some ways, the models inherent in these two metaphors are not all that different, it was none the less economics rather than biology that ultimately formed the basis of resource regulation in the fishery.

The second accomplishment of this work has been to demonstrate that the 'tragedy of the commons' metaphor is adequate neither to encompass nor explain the nature of the regulatory behaviour and property rights that occur within the Canadian East Coast inshore fishery. Indeed, just as the quotation from Greer that opens this chapter suggests, this metaphor quite literally suppressed many of the aspects of fishing activity and community regulation that were obviously well known to most everyone responsible for the implementation of the fisheries policy at the local level, and, in all likelihood, to those in more distant policy-making locales as well. Such 'tunnel vision,' in and of itself, would not have been a problem had the 'tragedy of the commons' metaphor been more or less consistent with what occurred on the fishing grounds. However, as our community studies have shown, in many instances, it was not, with the result that fishermen (and local-level Fisheries officers) were often placed in the position of having to implement regulatory policies that conflicted directly with functional and effective local 'work community' practices. In sum, the present study has looked at the intersection of competing metaphors about the nature of the fishery, and their implications, one conceiving of it as an open-access common-property resource, and the other, as essentially a community-based enterprise, consisting of local fishery work communities that regulate their own economic world.

Of course, since we completed our interviews with the fishermen of Bonavista and the other four communities, and during the time when we were writing this book, there have been new developments in Newfoundland's inshore fishery. Confronted with evidence of clearly declining catches of cod and other species, federal Fisheries Department officials recalculated their estimates of the 'maximum sustainable yield' of the various Newfoundland fisheries. By 1990 they were openly admitting previous errors and warning that overfishing was now threatening to deplete the species on which twenty thousand Newfoundland fishermen depended for some or all of their livelihood. Then, in 1991, the Newfoundland inshore cod fishery virtually collapsed. The dire predictions had proved all too true. In response, in April 1992, the federal government took the unprecedented step of closing the inshore cod and capelin fishery in the waters off much of Newfoundland for both the 1992 and the 1993 fishing season. Overnight, nearly twenty thousand fishermen (and all the communities that depended on them) were out of work in what could be the largest economic collapse in Canadian economic history, and perhaps even in world history. In short, the predictions of the fishermen of Charleston, King's Cove, Grates Cove, Fermeuse, Bonavista, and 850 other communities had proved correct. It is now questionable whether the Newfoundland inshore, community-based fishery has any future at all.

Under such circumstances, it is appropriate, in concluding this study, to ask what insight it can offer in the aftermath of the collapse and what guidelines it might provide for future fisheries policy, if and when the Newfoundland inshore fishery is reopened in 1994. One starting point is to consider the implications contained in this work with regard to the appropriate role of the state in future fisheries regulation.

In our judgment, several 'false conclusions' about the appropriate role of the state might be drawn. For example, advocates of the 'tragedy of the commons' perspective might well believe that the recent collapse of fish stocks beyond the level of sustainable yield constitutes, in itself, a validation of their position. After all, they have long suggested that this is the only possible outcome in situations in which there is open access to common property. Our work has surely demonstrated, however, that most inshore fishing had, in fact, been doubly regulated, by state and community institutional structures. It is therefore difficult to accept any argument claiming that the collapse of the inshore fishery was an inevitable outcome of under-regulation and open access. Furthermore, the collapse of the inshore fishery is more in keeping with the warnings of the majority of the inshore fishermen we interviewed than with the orientation and ideology of those responsible for regulating it. Most fishermen were adamant in their concern that fish stocks were being depleted and that overfishing by offshore draggers and inshore fishermen using new technologies was threatening their livelihood and community lifestyle.

Another erroneous conclusion that some might draw from this analysis is that the state should withdraw from the regulation of the Newfoundland inshore fishery. After all, we have emphasized that most communities already have their own regulatory structures, many of which were in place well before the state made any major attempt to systematically regulate Newfoundland's inshore fishery. Yet such a conclusion would ignore the reality that the inshore fishery of Newfoundland is contained within a much larger arena of national and international fishing activity. In view of the intense fishing activity of many foreign nations in the offshore sector, many of the fishermen we interviewed in fact called for more rather than less state regulatory activity. The state presence is absolutely necessary to protect the inshore fishery from being undermined by overfishing in other sectors. Without such state control, the inshore fishery has no chance for recovery or survival. It is clear as well that the state presence is also needed to locate and access markets for the fish caught by Newfoundland's inshore fishermen. With the possible exception of the union or the large fish plants, there is no other body capable of such an undertaking. It should be noted, however, as we demonstrated earlier, that the union and the fish plants are caught in a conflict of interest, in that they represent both the inshore and offshore fishing industries. Thus, the state, at both the federal and the provincial level, must accept the important role of ensuring that the interests of the smaller inshore operators are not overlooked in the activities of those agencies. Without such protection, the communal basis of the Newfoundland fishery would inevitably be permanently altered.

The state must take on other roles as well. If the East Coast fishery is to be adequately regulated, the state must, for a variety of reasons, also become involved in the fishing activity that occurs *within* the inshore fishery. For example, although each community may maintain some aspect of local community regulation, the fishing grounds of most communities overlap with those of other communities. In the communities studied here, the conflicts that such overlapping of community jurisdictions can produce were most clearly in evidence in Fermeuse. However, they were also present in Bonavista's domination of the fishing grounds of nearby communities. When all is said and done, it is the state that retains both the authority and the responsibility to regulate such disputes. Its authority comes from its right to maintain law and order, and its responsibility comes from the delegation of control over the fisheries to the federal government under the terms of the British North American Act of 1867 and of subsequent constitutional documents.

Still, it seems inadequate to limit the state, in either its federal or its provincial manifestation, to a policing function. In attempting to conceptualize the most appropriate role for the state with regard to the inshore fishery, two distinctions come to mind. The first is the distinction between *access* and *process*, to which we have referred previously in this analysis, and the second is the more Kantian distinction between *form* and *content*.

One of the distinguishing features of a common-property resource is the openness of access to it. It is this feature which, if left unregulated, would indeed set the stage for the 'tragedy of the commons,' and it is this aspect of property relations that local communities indeed find most difficult to manage. In doing so, they have the invidious task of making judgments about the legitimacy of claims made by various residents. By contrast, the state has had little difficulty controlling access to the inshore fishery through the implementation of licensing regulations. Control of access, then, should remain, in our opinion, the prerogative of the state. In Kantian terms, it should be the role of the state to control the 'form' in which the process of fishing must take place. The process of fishing within any community's fishing territory, however, has long been governed by community-based regulations and, we believe, should continue to be so governed. In other words, the local fishery 'work community' should be given primary responsibility for the 'content' of the fishery regulations that seek to govern the behaviour of those who have been granted access to its community fishing grounds by the state.

Such a declaration, however, does not address directly the question of how such regulation should proceed. A significant insight into that process is to be found in McGoodwin's recent work (1990) outlining what he perceives to be the causes and solutions of the 'crisis in the world's fisheries.' He lists 'seven basic strategies' employed by fisheries regulatory authorities worldwide for the purpose of either conserving fish stocks or deciding who should benefit from them (that is, allocating rights to harvest them) (163). Most of these strategies, such as closing off certain areas, setting seasons, or limiting entry, are measures that have been tried at one time or another in Canada's fisheries. However, the most intriguing of the regulatory methods described by McGoodwin is the 'institution of various forms of property rights.' As he puts it,

The main appeal of institution of a system of property rights is that fishers who consider certain fish stocks or fishing grounds their own property may voluntarily restrain their fishing effort and develop greater concerns for conservation and management ...

Social equity considerations are crucial when implementing this strategy, since to confer property rights is also to deny them to some who may have interests in the fishery. The problem is often best addressed by having recipients pay for the rights somehow, through rents, royalties, or whatever, to defray administrative and other costs. (177)

We would suggest that one solution to the problem of fisheries management at the community level would involve the state's formally allocating to each community the property rights in its local community fishing grounds that the members of that work community already believe they possess. If nothing else, this would recognize existing community regulatory practices and would serve to embed them in the context of the broader state regulations governing access to the fishing grounds. It would also establish explicitly that common-property rights are, in law as well as in deed, *community* property rights.

It is interesting to note that this sort of focus on community ownership has already been advocated, in a somewhat different context, by one Nova Scotia community. That instance occurred when the fishers of

Port Hawkesbury were informed in 1990 that the fish-processing plant in their community, owned by National Sea Products Ltd., was closing. This plant was involved primarily in the processing of fish caught not by inshore fishermen, but by deep-sea trawlers off northern Newfoundland and Labrador. National Sea Products announced that, with recent declines in these fish stocks, the number of fish-processing plants that had previously been required was no longer warranted and that it would be closing the Port Hawkesbury plant and shipping the fish to another plant for processing. In response, the fishers of Port Hawkesbury demanded that the government allocate to their community the quota of northern cod stock that had previously been processed by the plant in their community. They would then rent the quota to any fish-processing firm that guaranteed to have its ships unload the fish for processing at Port Hawkesbury.

Although this community proposal was clearly aimed at preventing the loss of local processing jobs, it reveals a great deal about the way in which fishers and fishing communities of all types see their relationship to the fish stocks on which their living is based. Whereas state and government regulations tend to allocate catch quotas of fish stocks in deep-sea waters to large fish-processing firms, fishers such as those of Port Hawkesbury may just as legitimately argue that those quotas should be allocated to the fishers themselves, as community property. Such allocation would, of course, give rise to a complete reorganization of the nature of property and power relations in the fishery. Instead of having large corporations control access to (and, hence, ownership of) the fish, the government would be granting such control to community-based groups of fishermen, each with a decided interest in protecting and maintaining the fish stocks on which their particular livelihood depended. Of course, such conservation would be possible only if the state regulatory agencies worked in close cooperation with the community groups, providing them with the best available data on maximum sustainable yield and on the productivity of certain types of fish-catching technologies.

Whether such proposals should be adopted as the basis on which property rights are allocated in the offshore fishery is not of direct concern here. This example has been introduced only to provide a perspective on the way in which the institutionalized structure of property relations in the Canadian East Coast fishery could and perhaps should be reorganized. As far as the inshore fishery is concerned, the time for considering options for radical change is clearly at hand. Indeed, in the face of the complete collapse of that fishery, there is an obvious need for a comprehensive reconsideration of the way in which property relations and other aspects of the inshore fishery are to be regulated in the future. To fail to take advantage of the two-year moratorium for this purpose would almost assuredly doom the fishery to repeated failure in years to come. If there is one lesson to be learned from this book, it is that there is a strong historical and practical basis for allocating the regulation of fishery property rights to the local fishing community. It is interesting to note that, in a recent review of some of the literature on the 'commons', Berkes and his colleagues (1989, 93) come to much the same conclusion:

In general, we propose that successful approaches to the commons dilemma are found in complementary and compatible relationships between the resource, the technology for its exploration, the property-rights regime and the large set of institutional arrangements. We also propose that combinations of property-rights regimes may in many cases work better than any single regime. The success of local level management, for example, often depends on its legitimization by central government ... In some cases, co-operative management arrangements (co-management) are needed, involving the sharing of power between government and local communities.

It is one thing, however, to argue that the fishery property rights of local communities need to be reorganized, and quite another to explain how this might best be done. Clearly, some of the solutions adopted in the past have not been adequate. For example, MacInnes and Davis (1990) have reviewed the role of committees of fishermen's representatives throughout Nova Scotia and have concluded that these attempts at involving the fishermen in an advisory capacity in local level regulation have failed to give them any effective decision-making function (17). These bodies may act as forums for discussion, but they give fishermen neither the opportunity to initiate plans and programs nor the power to decide which ones should be implemented.

What is needed, then, are new institutional structures that empower fishermen in both the design and the implementation of fisheries policy, at both a regional and a local level. Some sense of how such institutional arrangements might operate can be gleaned from a recent work by Ostrom (1992) dealing with the 'crafting' of institutions for the local government of irrigation systems. In that work, Ostrom sets out a number of design principles for such institutional arrangements, including the development of clear boundaries, the establishment of mechanisms to monitor costs and benefits on an on-going basis, and the formation of conflict-resolution mechanisms (61–76). Within the fishery it would also be necessary to develop ways of evaluating the impact that technological changes in both the means and the technology of production would be likely to have, over time, on the opportunity chances of others.

We would argue, however, that a focus on the institutional arrangements *per se* fails to comprehend fully the power structure within which these institutions are found. For institutional rearrangements to be successful, there must be some understanding of, and preparation for, the ways in which they are likely to affect the existing balance of power. In the Newfoundland East Coast fishery, the major institutions holding power include the federal and provincial governments and their regulatory agencies, the unions, and the major fish-processing firms.

If property rights in the inshore fishery are to be formally lodged with the work communities themselves, those who fish from these communities must be given real power to decide on allowable catch allocations, appropriate use of gear, licensing regulations, and a variety of other dimensions of fishing practice. In other words, they must have more than a simple advisory role. Such an empowerment of fishermen will, of necessity, alter the power balance that currently exists in the fishery. For example, it may well be that the fishermen should be granted an allocation of catch that they can then, on a community-by-community basis, sell or lease to various fish-processing companies. If such an arrangement were implemented, the balance of power between fishermen and fish-processing firms would clearly change.

Similarly, a reallocation of control over the fishery on a community-by-community basis would be likely to affect the power and role of inshore fishermen in relation to their union. Since the union already has a fishermen's committee in each community, it would seem likely that such groups would seek to represent their communities' interests, and, in some instances, this might well be appropriate. To the extent that the union could consolidate the various community interests into one block, it would see its power increase substantially *vis-à-vis* the fish-processing firms. However, this possible outcome might also result in community's fishers having less real power than they had previously. Their power would simply have been transferred to a Leviathan-like body that would then act in their supposed interest. The real possibility of such a scenario suggests an appropriate new role for provincial fisheries officials – namely, ensuring that each community's new regulatory power over its fishing property is not simply supplanted by larger economic forces and institutional structures.

All of this, of course, is speculation. Before any reallocation of fishery property rights to inshore fishermen can occur, careful study and attention must be given to the way in which communities may actually use these rights and to the question of who in a community should have the power to do so. However, the possibility that such changes may significantly alter existing power relationships within the fishery should not be permitted to impede efforts to bring them about. At the end of chapter 7 we noted how policy makers whose job it was to regulate the fishery on behalf of fishermen had failed those whose interests they were supposedly most intent on protecting. There is now a clear need to have the state work more directly *with* inshore fishermen, not *for* them – and most definitely not *against* them, as has occurred in the past – in creating new regulatory structures and new institutional arrangements that will ensure the survival of the inshore fishery. No other alternative is acceptable.

Notes

Chapter 1

1 The term 'inshore fishery' is generally used in reference to small boats (that is, those of less than about thirty-five feet) that operate within sight of the coast. The term 'nearshore fishery' is generally used in reference to longliners of thirty-five feet and more that have the potential to operate outside the range generally used by the inshore boats. However, as will be discussed later in this book, the two types of boats often come into conflict because many longliner operators choose to operate in inshore waters. Consequently, the primary distinction between the two 'fisheries' lies in the different technologies that they use to prosecute the fishery, and has little to do with the space in which they operate. Much more will be said about this later in the book. However, both terms are used to distinguish the small-boat, community-based, local fishery from the 'offshore' fishery carried out by the considerably larger trawlers and draggers that generally fish the 'banks' that lie well outside the range of the inshore and nearshore boats. But even here there are exceptions: Many of the crab-fishing boats, which are longliners, actually operate considerably farther from shore than many of the much larger trawlers and draggers. Once again, the distinction between these types of fisheries is seen to rest more on the technology employed on the boat than on the size of the boat or the location or space in which the fishery takes place.

2 For a more detailed discussion of what is implied by the term *technologies of production* in the context of the Canadian East Coast fishery, see 'Class Interests and the Role of the State in Canada's East Coast Fishery,' in Matthews 1983, 194–8.

3 A note on terminology is necessary here. The term *fishermen* is sexist in that it refers, literally, only to men who fish. With the growing awareness of sexist language, those who write about the fishery have sought some alternative. One solution is to employ the generic term *fishers* in reference to both men and women who engage in the fishery. But this solution is problematic, because the majority of those who engage in the fishery have historically been, and continue to be, men, and they refer to themselves as 'fishermen,' not 'fishers.' (Another solution is to use the terms *fishermen* and *fisherwomen*, but this option has not found favour among writers, perhaps because it is cumbersome.) In this book, we have chosen to retain *fishermen* as a generic term, in reference to all who engage in the fishery. In some instances, we have also used the term *fishers* generically; for the most part, however, we have used that term in reference to women who engage in the fishery.

4 The policy against giving out names of union members had been voted into effect at an annual union meeting in order to prevent union membership lists from being sold for commercial purposes. The union official was willing to bring my request before the next annual meeting, but that was nearly a year away.

5 Readers who wish to obtain a copy of the interview schedule used in this study may do so by writing to the author, care of the Department of Sociology, McMaster University, Hamilton, Ontario, Canada, L8S 4M4.

Chapter 2

1 St John's is located more than 500 nautical miles east of Halifax and less than 1900 nautical miles from Ireland.

2 This was achieved largely through the expedient of refusing to make payments on the outstanding foreign debt.

3 The Confederation agreement also included the payment by Canada of all of Newfoundland's outstanding debt, thus freeing the society from the financial burden that had driven it into bankruptcy.

4 Between 1951 and 1981 the population of St John's and its surrounding area increased from 67,749 to 108,253, while that of Newfoundland as a whole increased from 361,819 to 567,681 (complied from data contained in Newfoundland and Labrador 1985, tables A–1, A–9).

5 All labour-force statistics given here are calculated on the basis of data contained in Newfoundland and Labrador, Department of Finance 1985, table C–5.

6 Different statistical agencies provide quite different estimates of the

number of fishermen in Newfoundland. This problem stems partly from the seasonality of the inshore fishery. For example, few respondents to a survey taken in mid-winter are likely to indicate current employment in the fishery. Other differences stem from differences in the definition of the term *fisherman*. Some studies appear to include only those whose income comes exclusively from fishing; others include only those who hold a full-time fishing licence; while still others include only those who claim unemployment insurance from fishing. The 1985 edition of the *Historical Statistics of Newfoundland and Labrador*, from which the labour-force statistics used in this discussion are drawn, indicates that in 1951 there were 18,420 fishermen in Newfoundland, constituting 17.3 per cent of the labour force; that by 1971 there were only 6810 fishermen, constituting 4.6 per cent of the labour force; and that by 1981 the number of fishermen had increased to 9895 but the labour force had expanded at an even greater rate, with the result that fishermen still represented only 4.5 per cent of the labour force. Information from other sources however, suggests that these figures seriously underestimate the number of active full- and part-time fishermen in Newfoundland.

7 At least 137 communities were resettled under the program between 1965 and 1972 (see Matthews 1987, 2441).

8 In most of the province, the season for salmon and lobster is May and June. Although salmon and lobster have traditionally provided a welcome source of cash, stocks have never been adequate to support a major fishery.

9 A handline, or jigger, consists of a line with a lead weight on it, which is 'jigged' up and down through the water without bait. The fish are attracted to the movement of the jigger and are hooked, one at a time, when they swim too close, to investigate. Trawling is a more efficient method of fishing. The trawl, consisting of several hundred feet of line with baited hooks attached at intervals, is submerged in the water in areas where the cod are known to 'run.' The disadvantage of the trawl is that the fish sometimes die after being hooked, and may decay in the water for several hours or days before they are brought ashore. Cod nets, like trawls, are submerged in water in areas where the fish are presumed to run. The fish swim into them, and their gills become entangled in the netting. Unable to breathe, the fish quickly drown. Thus, with this method, there is even more likelihood of decay before harvesting. A cod trap is essentially a very large net square with a net bottom. On one side is a slit with a long net leader. The trap is submerged in the water with the leader facing the direction in which the fish are most likely to swim. As the twine used in

making a trap is quite large, the fish tend to see leader and attempt to swim around it. Instead, they swim through the opening into the trap, where they mill about, unable to find their way back out. Hundreds of fish can be caught in this manner and remain alive until the trap is raised to the surface, where they can be scooped out with dip nets.

10 Both McCay (1978, 414; 1979, 167) and Martin (1979) describe this process. It will be a major consideration in our community analysis in chapter 5, 6, and 7.

11 For example, trawls are easier to haul from smaller boats with relatively low sides, while the stability and size of larger boats are advantageous for hauling traps. In addition, it usually requires two boats, one on each side, to haul a trap.

12 It should be pointed out, however, that all the men involved in the harvesting unit can be viewed as appropriating the value of the labour added by the women and children involved in the processing unit.

13 Consequently, they also still bear some of the burden of risk in a way that is not characteristic of most wage labour.

14 For a discussion of the factors contributing to these developments, see Sinclair 1986.

15 The minimum number of weeks of employment required varies by region of the province, depending on the level of unemployment in each region.

Chapter 3

1 For an in-depth discussion of the characteristics of 'goal-value systems,' see Matthews 1975 and 1983, 137–47.

2 A more biologically oriented review of Canadian fisheries-management regulations is found in Needler 1979. Because Needler, a leading Canadian fishery biologist employed by the federal government, was actively involved in the development of many of these policies, his history of their development is a particularly important documentary source.

3 Although they are omitted from McCorquodale's three periods, the years from 1970 to 1977 are examined in her discussion under the heading 'Fisheries Policy, 1950 to 1970.'

4 With regulatory jurisdiction over fishing within 200 miles of its coastline, Canada could limit the extent of fishing by foreign fishing fleets on the various banks situated about 50 to 100 miles offshore. The greatest impact of Canada's new power was on its offshore fleet, and this had an indirect impact on the inshore and nearshore fishery as well. With the acquisition of control over the offshore fish stocks, Canada's fishery planners clearly

came to believe that they were in a position to 'rationalize' all aspects of the fishing industry, including the inshore and processing sectors. As part of this thrust towards 'rationalization,' Canada developed regulations designed to increase the fish stocks that had been depleted by the overfishing of offshore vessels.

5 We will examine the nature of common property in more detail in chapter 4. However, it is important to note here that most writers have tended to equate common property with open access. As we shall see in chapter 4, the two are analytically and practically separable, and this has major implications for the development of fisheries policy.

6 This is particularly evident in Needler 1979, where the author demonstrates his own acceptance of the common-property perspective as the proper basis for fisheries management. (Our discussion of this point is not intended to suggest in any way that fishery biologists ceased to play a significant role in their own right in the resource management of Canada's East Coast fishery.)

7 Economists also had very legitimate claims to expertise in matters pertaining to the production and marketing of fish, and this further strengthened their ability to influence the development of fishery policies.

8 See Scott 1979 for a listing and description of the most important of these works.

9 Despite such assertions of universality and inexorability, much of our discussion of the nature of common property in chapter 4 will centre on situations and conditions where the common-property character of a resource does not necessarily lead to its exploitation and depletion.

10 It is interesting to note how accurately Crutchfield's analysis predicted the plight of many Newfoundland longliner operators in the wake of licensing. As we shall see from our community studies, many of them have faced bankruptcy because they were unable to obtain the range of species licences that would have enabled them to earn sufficient income to cover the financing and operating costs of their boats.

11 The discussion of the West Coast experience that follows is a synopsis of several works on limited-entry regulation of the West Coast fishery during this period, including Fraser 1979, Hayward 1981, Mundt 1975, Pearse and Wilen 1979, and Retting and Ginter 1978.

12 Such code words are commonly referred to as 'jargon' by those who do not subscribe to the particular paradigm.

13 As the name suggests, groundfish are species of fish that swim near the ocean floor rather than near the surface.

14 Copes (1983, 16–17) documents that between 1974 and 1980 the number

of registered fishermen in Newfoundland rose from 15,351 to 35,080. It is likely that much of this increase in the number of registered fishermen was attributable to registrations by people who were attempting to protect themselves from being refused access to the fishery in the future. Newfoundlanders were generally well aware that such a registry could be used in future as a record of past participation and as a means of determining who were the legitimate, or 'bona fide,' fishermen. Copes's data tend to support this interpretation, in that he estimates that in 1980 there were only 21,297 persons for whom the fishery was an occupation (1983: 16).

15 Levelton was commissioned to write the report in 1978, at which time the terms of reference required that he 'provide to the minister periodic progress reports and a final report by April 15th, 1979' (Levelton 1981). Although the report was presumably available to the minister at that time, it was not published until 1981.

16 In order to keep the historical chronology straight, it should be noted that Sinclair (1981, 22) indicates that LeBlanc first announced the inauguration of limited-access licensing in an address to the Newfoundland Fishermen, Food and Allied Workers Union in November 1980. Although LeBlanc did announce the program there and elaborate on the distinction between full- and part-time fishermen, the *first* announcement was actually made at the Memramcook meeting, a month earlier.

17 Ironically, it was the same dilemma that had brought down the earlier attempt to 'rationalize' the fishery, through the centralization and resettlement of the population. Though the 1978 White Paper explicitly rejected any such strategy, its authors appear to have been unaware that the same issues were involved in licensing and limited access. (See Novak 1980 for a detailed discussion of this link.)

18 The committee was established by the president of Memorial University of Newfoundland, apparently because of his own concerns about the impact of licensing policy on Newfoundland society. It consisted entirely of faculty members of Memorial University, many of whom had previous research experience related to rural Newfoundland fishing communities.

19 First Ministers Conferences are meetings that bring together all the provincial premiers and the prime minister of Canada.

20 A 'fish glut' occurs when more fresh fish is caught in a given area than the processing plant in that area can process. Fish gluts are quite frequent when the fish 'strike in' in pursuit of capelin or the other smaller fish on which they feed.

21 It is significant that, even at this early stage, the union was using the term *bona fide fishermen*, the federal government's code word for full-time

fishermen, whom it wished to support at the expense of part-time fisher-men. This indicates how closely aligned the union and federal-government positions were on the subject of limited access.

Chapter 4

1 It is worth noting, however, the recent attempts by some conservative lobby groups in Canada to have the individual's 'right to own property' enshrined in the Constitution. Should such efforts ultimately prove successful, this could constitute a limitation on the power of the state with regard to the enforcement of property rights.
2 This statement applies whether one defines common property in the usual sense of the term, as a resource to which access is totally open, or in the sense we have adapted in this discussion, as 'community common property.'
3 The insurance value to which we refer differs slightly from simply protecting against the possibility of an absence of fish in some locations in particular years. The role of property 'insurance' in any circumstance is to protect property holders from a sudden loss so severe that it might offset any gains made in previous years and leave the property holders so short of resources as to make it impossible for them to continue their traditional way of earning a living. While it is quite likely that some fishing areas are more profitable than others, there is no guarantee that they will be more profitable in any given year or years. And although it might well be in the interest of some individuals to hold on to such 'prime' locations if they can, the possibility that they could be 'wiped out' by a series of successive crop failures is sufficient incentive to keep them committed to a common-property system. This is particularly true in the case of the inshore fishery, where few operators have the personal resources to offset a major loss.
4 MacPherson and Marchak adopt essentially a structural perspective, as do most economists who adopt the neoclassical perspective. Similarly, both the new institutional economics and public-choice theory are also con-cerned primarily with the social-structural aspects of common-property relationships. This is not to say that such sources do not, on occasion, examine the nature of values and meanings or that they ignore individual actions in their analysis, but only that they do not take as their starting point the way the individual actions of affected actors are shaped by their values and the meanings they construct.
5 One of the problems with the cost–benefit approach is that it is remarkably similar to the theory of motivation that underlies the 'tragedy of the

commons' position, which it opposes. Both positions assume that we all act on the basis of economically rational motives or, at least, on what we perceive to be the most economically rational action to take. This view fails to take into account a plethora of other aspects and bases of motivation, including commitment to location, lifestyle, and family. In an earlier study (Matthews 1976), this writer differentiated between considerations of 'economic viability' and 'social viability,' which enter into the decision-making process at both an individual and a community level. That work left no doubt that the 'costs and benefits' approach is largely unable to explain the complex range of considerations and definitions taken into account by the actors in determining their choice of action.

Chapter 5

1 It will be argued later in the chapter that, at least with respect to the inshore fishery, devising strategies of work and devising strategies of earning a living from that work are two different separate sets of activities.
2 According to King's Cove residents, during a brief period at the start of the fishing season, all fishermen in the community hold special herring licences that allow them to fish for herring solely for use as bait in lobster pots.

Chapter 6

1 Indeed, with the high interest rates of recent years, many skippers have faced bank foreclosures, losing both their boats and the capital they invested in them.
2 Some crews have an additional third trap, which they use essentially as a backup or insurance factor. Should holes develop in one of the two main traps (as is frequently caused by whales or high winds), it can be replaced by the backup trap. This allows the crew to do repairs without losing much fishing time during the crucial six-week period when the traps are effective.
3 Regulations do permit a father to pass his protected-species licence on to his son if the son has always fished with the father. However, if the son fishes by himself or with another crew, he is not eligible to inherit the licence.
4 Although only seven people responded to the question about the capability of leaders, all were positive. From the more general comments of the other interviewees, it would appear that they did not think of their community as

having leaders because the community had no elected community council at the time that they were interviewed.

5 Not to be confused with the southern coast of Newfoundland, the Southern Shore runs in a vaguely north-south direction. Its name comes from its being the shoreline that is south of St John's.

6 However, Calvert found the climate harsh and, after two years, returned to England where he was named Lord Baltimore and awarded a patent to what is now Maryland. He died before he could establish a colony there. However, his son subsequently established a colony at Baltimore.

7 This presumably explains the confusion among some respondents about whether the community had ten or eleven trap berths.

8 Gill-netting was not practised in 1927, when this regulation was introduced.

9 It should be noted that, according to actual data on licensing composition, the full-time–to–part-time ratio is roughly the same in both communities. In social life, however, factual reality is often not as important as the participants' perception of reality. There can be no doubt that the fishermen of Fermeuse genuinely believe that the majority of Renews fishermen are part-timers, and base their social activities on that assumption.

Chapter 7

1 For those interested in the origins of other families in Bonavista, Seary (1977) records the following information. By 1792 the following family names appear in records about the community: Baker, Mouland, Powers, Short, Oldford, Phillips, Mifflin, Dyke, Collins, Little, Fisher, Hailey, Hayward, Hobbes, Burton, Shirran, Stag, Street, White, and Waye. By the turn of the century, these were joined by forebears of the following families: Groves (1793), Durdle (1776), Randell (1796), Squires (1798), Russell (1799). In the early years of the nineteenth century, there are records of other family names, including Fifield (1800), Shi(e)lly and Linthorne (1801), Templeman (1803), Ayles and Sweetland (1804), Butler and Hancock (1805), Holloway and Hunt (1808), Harris (1810), Matcham (1811), O'Connell (1815), Sellars (1817), Tremblett (1819), Marsh and Hampton (1823), Sharpe and Soper (1826), Gibbs (1827), Wiffen (1828), Reid (1830), Hodder (1854), and Hugh and William Faulkner (1871).

2 In Bonavista we did not sample protected-species licence holders separately.

3 Some fishermen implied that the high insurance rates were largely the consequence of a number of unprofitable boats' catching fire in mysterious circumstances.

4 We encountered two cases in Bonavista in which the crab-boat licence was held not by an individual but by a crab-processing plant in the community. In each of these cases, the boat was also owned by the processing company and the skipper was a paid employee of the company. The implications of such arrangements will be discussed later in the section.

References

Acheson, James M. 1975. 'The Lobster Fiefs: Economic and Ecological Effects of Territoriality in the Maine Lobster Industry.' *Human Ecology* 3: 2, 183–207
– 1987. 'The Lobster Fiefs Revisited: Economic and Ecological Effects of Territoriality in the Maine Lobster Industry.' In McCay and Acheson (1987), 37–65
Andersen, Raoul, and Cato Wadel. 1972. *North Atlantic Fishermen: Anthropological Essays on Modern Fishing*. Newfoundland Social and Economic Papers no. 5. St John's: Institute of Social and Economic Research, Memorial University of Newfoundland
Barrett, L. Gene. 1981. 'The State and Capital in the Fishing Industry: The Case of Nova Scotia.' Paper presented at the annual meeting of the Canadian Political Science Association, Dalhousie University, Halifax, NS, 27 May
Berger, Peter L., and Thomas Luckman. 1967. *The Social Construction of Reality: A Treatise on the Sociology of Knowledge*. Garden City, NY: Doubleday and Co.
Berkes, Fikret. 1977. 'Fishery Resource Use in a Subarctic Indian Community.' *Human Ecology* 5: 4, 289–307
– 1983. 'A critique of the "Tragedy of the Commons" paradigm.' Paper presented at the Natural Management Systems Symposium (I-A181), Ninth International Congress of Anthropological and Enthnographic Sciences, Quebec City
– 1985. 'The Common Property Resource Problem and the Creation of Limited Property Rights.' *Human Ecology* 13: 2, 188–208
– 1986. 'Marine Inshore Fishery Management in Turkey.' In National Research Council, Office of International Affairs, Board on Science and

Technology for International Development, ed., *Proceedings of the Conference on Common Property Resource Management*, 63–84. Washington, DC: National Academy Press

– 1987. 'Common Property Resource Management and Cree Indian Fisheries in Subarctic Canada.' In McCay and Acheson 1987, 66–91

Berkes, Fikret; David Feeny; B.J. McCay; and J.M. Acheson. 1989. 'The Benefits of the Commons.' *Nature* 340 13 July, 91–93

Blomquist, William, and Elinor Ostrom. 1985. 'Institutional Capacity and the Resolution of a Commons Dilemma.' *Policy Studies Review* 5: 383–95

Blumer, Herbert. 1948. 'Public Opinion and Public Opinion Polling.' *American Sociological Review* 13: 542–4

Brox, Ottar. 1972. *Newfoundland Fishermen in the Age of Industry: A Sociology of Economic Dualism*. Newfoundland Social and Economic Studies no. 9. St John's: Institute of Social and Economic Research, Memorial University of Newfoundland

Canada. Department of Fisheries and Oceans. 1979. 'Freeze on Inshore Groundfish Fishing Licences Partially Lifted.' *News Release: Comuniqué*, no. NR-HQ-79-30E. Ottawa.

– Environment Canada. Fisheries and Marine Service. 1976. *Policy for Canada's Commercial Fisheries*. Ottawa: Supply and Services Canada

– House of Commons. 1973. *Official Minutes (Hansard)*. 14 November. Ottawa: Supply and Services Canada

– Task Force on Atlantic Fisheries. 1982. *Navigating Troubled Waters: A New Policy for the Atlantic Fisheries. Report of the Task Force on Atlantic Fisheries*. Chaired by Michael J.L. Kirby. Ottawa

Chiaramonte, Louis. 1971. *Craftsman–Client Contracts: Interpersonal Relations in a Newfoundland Fishing Community*. Newfoundland Social and Economic Studies no. 10. St John's: Institute of Social and Economic Research, Memorial University of Newfoundland

Committee on Federal Licensing Policy. 1974. *Report of the Committee on Federal Licensing Policy and Its Implications for the Newfoundland Fisheries*. St John's: Institute of Social and Economic Research, Memorial University of Newfoundland

Connelly, M. Patricia, and Martha MacDonald. 1983. 'Women's Work: Domestic and Wage Labour in a Nova Scotia Community.' *Studies in Political Economy* 10 (Winter), 45–62

Copes, Parzival. 1980. 'The Evolution of Marine Fisheries Policy in Canada.' In Peter N. Nemetz, ed., *Resource Policy: International Perspectives*, 125–48. Montreal: Institute for Research on Public Policy.

– 1983. 'Fisheries Management on Canada's Atlantic Coast: Economic Factors

and Socio-Political Constraints.' *Canadian Journal of Regional Science* 6: 1,
1–32

Crutchfield, J.A. 1979. 'Economic and Social Implications of the Main Policy
Alternatives for Controlling Fishing Effect.' *Journal of the Fisheries Research
Board of Canada* 36: 742–53

Davis, Anthony. 1984. 'Property Rights and Access Management in the Small
Boat Fishery: A Case Study from Southwest Nova Scotia.' In Cynthia
Lamson and Arthur J. Hanson, eds., *Atlantic Fisheries and Coastal Commu-
nities: Fisheries Decision-Making Case Studies*, 133–64. Halifax: Ocean
Studies Programme, Dalhousie University

Davis, Anthony, and Leonard Kasdan. 1985. 'Bankrupt Government Policies
and Belligerent Fishermen's Responses: Dependency and Conflict in the
Southwest Nova Scotia Small Boat Fisheries.' *Journal of Canadian Studies*
19: 1, 108–24

Davis, Anthony, and Victor Thiessen. 1986. 'Public Policy and Social Control
in the Fisheries.' Paper prepared for the International Working Seminar on
Social Research and Public Policy Formation in the Fisheries: Norwegian
and Atlantic Canadian Experiences, University of Tromsø, Norway,
23–27 June

Denzin, Norman K. 1970. *The Research Act: A Theoretical Introduction to
Sociological Methods.* Chicago: Aldine

Draper, Dianne. 1981. 'Ocean Exploitation: Efficiency and Equity Questions in
Fisheries Management.' In Bruce Mitchell and W.R. Derrick Sewell, eds.,
Canadian Resource Policies: Problems and Prospects, 109–50. Toronto:
Methuen

Eckberg, Douglas Lee, and Lester Hill, Jr. 1979. 'The Paradigm Concept and
Sociology.' *American Sociological Review* 44: 925–37

Fairley, Bryant O. 1985. 'The Struggle for Capitalism in the Fishing Industry
in Newfoundland.' *Studies in Political Economy* 17: 33–69

Faris, James C. 1966. *Cat Harbour: A Newfoundland Fishing Settlement.*
Newfoundland Social and Economic Studies no. 3. St. John's: Institute of
Social and Economic Research, Memorial University of Newfoundland

Fernandez, J.W. 1987. 'The Call to the Commons: Decline and Recommitment
in Austria, Spain.' In McCay and Acheson 1987, 266–89

Fine, Gary Alan. 1979. 'Small Groups and Culture Creation.' *American
Sociological Review* 44: 733–45

– 1983. *Shared Fantasy: Role-Playing Games as Social Worlds.* Chicago:
University of Chicago Press

Finestone, Melvin M. 1967. *Brothers and Rivals: Patrilocality in Savage Cove.*
Newfoundland Social and Economic Studies no. 5. St John's: Institute of

Social and Economic Research, Memorial University of Newfoundland

Fraser, G. Alex. 1979. 'Limited Entry: Experience of the British Columbia Salmon Fishery.' *Journal of the Fisheries Research Board of Canada* 36: 757–62

Giddens, Anthony. 1979. *Central Problems in Social Theory: Action, Structure and Contradiction in Social Analysis.* Berkeley: University of California Press

– 1984. *The Constitution of Society: Outline of the Theory of Structuration.* Berkeley: University of California Press

Gordon, H. Scott. 1954. 'The Economic Theory of a Common-Property Resource: The Fishery.' *Journal of Political Economy* 62:2, 124–42.

Great Britain. 1933. *Report of the Royal Commission on Newfoundland.* London: King's Printer

Gunder Frank, André. 1969. *Latin America: Underdevelopment or Revolution?* London: Modern Reader Paperback, Monthly Review Press

Handcock, W. Gordon. 1977. 'English Migration to Newfoundland.' In John J. Mannion, ed., *The Peopling of Newfoundland: Essays in Historical Geography,* 15–48. Newfoundland Social and Economic Papers no. 8 St John's: Institute of Social and Economic Research, Memorial University of Newfoundland

Hanson, Arthur J., and Cynthia Lamson. 1984. 'Fisheries Decision Making in Canada.' In Cynthia Lamson and Arthur J. Hanson, eds., *Atlantic Fisheries and Coastal Communities: Fisheries Decision-Making Case Studies,* 1–15. Halifax: Ocean Studies Programme, Dalhousie University

Hardin, Garrett. 1968. 'The Tragedy of the Commons.' *Science* 162: 1243–8

– 1977. 'What Marx Missed.' In Garrett Hardin and John Baden, eds. , *Managing the Commons,* 3–7. San Francisco: W.H. Freeman

Hayward, Brian. 1981. 'The B.C. Salmon Fishery: A Consideration of the Effects of Licensing.' *B.C. Studies* 50: 39–51

Hobbes, Thomas. [1651] 1962. *Leviathan.* Reprint. London: Collier Macmillan

House, J. Douglas. 1986. 'Canadian Fisheries Policies and Troubled Newfoundland Communities.' Paper prepared for the International Working Seminar on Social Research and Public Policy Formation in the Fisheries: Norwegian and Atlantic Canadian Experiences, University of Tromsø, Norway, 23–27 June

House Royal Commission. *See* Newfoundland and Labrador. Royal Commission on Employment and Unemployment

Iverson, Noel, and D. Ralph Matthews. 1968. *Communities in Decline: An Examination of Household Resettlement in Newfoundland.* Newfoundland Social and Economic Studies no. 6. St John's: Institute of Social and Economic Research, Memorial University of Newfoundland

Judah, Charles Burnet. 1933. *The North American Fisheries and British Policy to 1713*. Illinois Studies in the Social Sciences. Urbana, IL: University of Illinois Press

Kirby Task Force. *See* Canada. Task Force on Atlantic Fisheries

Kuhn, Thomas S. 1970. *The Structure of Scientific Revolutions*. Chicago: University of Chicago Press

LeBlanc, Roméo. 1978a. 'Notes for a Speech by the Hon. Roméo LeBlanc to the Fishery Ministers of the Atlantic Region.' Moncton. NB, 10 November. Transcript provided by the Department of Fisheries and Oceans, Ottawa

– 1978b. 'Notes for Remarks by the Hon. Roméo LeBlanc, Minister of Fisheries at the Federal–Provincial Conference of First Ministers on the Economy.' Ottawa, 29 November. Transcript provided by the Department of Fisheries and Oceans, Ottawa

– 1980a. 'An Address by the Hon. Roméo LeBlanc, Minister of Fisheries and Oceans, at the Gulf Ground Fish Seminar.' Memramcook, NB, 25 September. Transcript provided by the Department of Fisheries and Oceans, Ottawa

– 1980b. 'Notes for a Speech by the Hon. Roméo LeBlanc, Minister of Fisheries and Oceans, at the 50th Anniversary Meeting of the United Maritime Fishermen.' Moncton, NB, 19 March. Transcript provided by the Department of Fisheries and Oceans, Ottawa

Levelton, C.R. 1981. *Toward an Atlantic Coast Commercial Fisheries Licensing System: A Report Prepared for the Department of Fisheries and Oceans*. Ottawa: Department of Fisheries and Oceans

Lloyd, W.F. [1933] 1977. 'On the Checks to Population.' Reprinted in Garrett Hardin and John Baden, eds., *Managing the Commons*, 8–15. San Francisco: W.H. Freeman

Locke, John. [1689] 1952. *The Second Treatise of Government*. Indianapolis: Bobbs-Merrill

Long, Norman. 1977. *An Introduction to the Sociology of Rural Development*. London: Tavistock

Lyman, Stanford M., and Marvin B. Scott. 1970. *A Sociology of the Absurd*. New York: Appleton-Century-Crofts

McCay, Bonnie J. 1978. 'Systems Ecology, People Ecology, and the Anthropology of Fishing Communities.' *Human Ecology* 6: 4, 397–422

– 1979. 'Fish Is Scarce': Fisheries Modernization on Fogo Island, Newfoundland.' In Andersen 1979, 155–88

– 1987. 'The Culture of the Commoners: Historical Observations on Old and New World Fisheries.' In McCay and Acheson 1987, 195–216

McCay, Bonnie J., and James M. Acheson, eds. 1987. *The Question of the*

Commons: The Culture and Ecology of Communal Resources. Tucson: University of Arizona Press.

McCorquodale, Susan. 1983. 'The Management of a Common Property Resource: Fisheries Policy in Atlantic Canada.' In Michael M. Atkinson and Marsha A. Chandler, eds., *The Politics of Canadian Public Policy*, 151–71. Toronto: University of Toronto Press

MacDonald, Martha, and M. Patricia Connelly. 1986a. 'Household Labour Force Activity in Rural Communities.' Paper presented at the annual meeting of the Canadian Economics Association, Winnipeg, 29 May

– 1986b. 'Workers, Households, Community: A Case Study of "Restructuring" in the Nova Scotia Fishery.' Paper presented at the International Working Seminar on Social Research and Public Policy Formation in the Fisheries: Norwegian and Atlantic Canadian Experiences, University of Tromsø, Norway, 23–27 June

MacDonald, R.D.S. 1984. 'Canadian Fisheries Policy and the Development of Atlantic Coast Groundfisheries Management.' In Cynthia Lamson and Arthur J. Hanson, eds. , *Atlantic Fisheries and Coastal Communities: Fisheries Decision-Making Case Studies*, 15–76. Halifax: Dalhousie Ocean Studies Programme

McGoodwin, James R. 1990. *Crisis in the World's Fisheries: People, Problems, and Policies.* Stanford, CA: Stanford University Press

MacInnes, Daniel, and Anthony Davis. 1990. 'Representational Management or Management of Representation?: The Place of Fishers in Atlantic Canadian Fisheries Management.' Department of Sociology and Anthropology, St Francis Xavier University. Photocopy

MacKenzie, W.C. 1979. 'Rational Fishery Management in a Depressed Region: The Atlantic Groundfishery.' *Journal of the Fisheries Research Board of Canada* 36: 811–54

Macpherson, Alan C. 1977. 'A Modal Sequence in the Peopling of Central Bonavista Bay.' In John J. Mannion, ed., *The Peopling of Newfoundland: Essays in Historical Geography.* Newfoundland Social and Economic Papers no. 8. St John's: Institute of Social and Economic Research, Memorial University of Newfoundland

MacPherson, C.B. 1978. 'The Meaning of Property.' In C.B. MacPherson, ed., *Property: Mainstream and Critical Positions*, 1–13. Toronto: University of Toronto Press

Marchak, M. Patricia. 1987. 'What Happens When Common Property Becomes Uncommon?' Department of Anthropology and Sociology, University of British Columbia. Photocopy

Martin, Kent O. 1973. 'The law in St John's says ...' MA thesis, Memorial
    University of Newfoundland
– 1979. 'Play by the Rules or Don't Play at All: Space Division and Resource
    Allocation in a Rural Newfoundland Fishing Community.' In Andersen 1979,
    277–98
Matthews, Keith. 1968. *The History of the Newfoundland West Country Fishery.*
    PhD diss., Oxford University
Matthews, Ralph. 1970. *Communities in Transition: An Examination of
    Government Initiated Community Migration in Rural Newfoundland.* PhD
    diss., University of Minnesota
– 1975. 'Ethical Issues in Policy Research.' *Canadian Public Policy* 1: 2,
    204–216
– 1976. *There's No Better Place Than Here: Social Change in Three Newfound-
    land Communities.* Toronto: Peter Martin Associates. Reprint 1982. Toronto:
    Irwin Publishing
– 1979. 'The Smallwood Legacy: The Development of Underdevelopment in
    Newfoundland, 1949–1972.' *Journal of Canadian Studies* 13: 89–108
– 1980. 'Class Interests and the Role of the State in the Development of
    Canada's East Coast Fishery.' *Canadian Issues: Journal of the Association for
    Canadian Studies* 3: 1, 115–24
– 1983. *The Creation of Regional Dependency.* Toronto: University of Toronto
    Press
– 1987. 'The Outport Breakup.' *Horizon Canada* 9: 102, 2438–43
– 1988. 'Federal Licensing Policies for the Atlantic Inshore Fishery and Their
    Implementation in Newfoundland, 1973–1981.' *Acadiensis: Journal of the
    History of the Atlantic Region* 17: 2, 83–108
Matthews, Ralph, and John Phyne. 1988. 'Regulating the Newfoundland In-
    shore Fishery: Traditional Values versus State Control in the Regulation of a
    Common Property Resource.' *Journal of Canadian Studies* 23: 1 and 2, 158–76
Moloney, David G., and Peter H. Pearse. 1979. 'Quantitative Rights as an
    Instrument for Regulating Commercial Fisheries.' *Journal of the Fisheries
    Research Board of Canada* 36: 859–66
Moores, Frank D. 1978. 'Fisheries in the Future.' Paper presented at the First
    Ministers Conference, Ottawa, 13–15. February. Transcript provided by
    the Office of the Premier, Government of Newfoundland and Labrador,
    St John's
Mundt, J. Carl, ed. 1975. *Limited Entry into the Commercial Fisheries.* Seattle:
    Institute for Marine Studies, University of Washington
Needler, A.W.H. 1979. 'Evolution of Canadian Fisheries Management towards

Economic Rationalization.' *Journal of the Fisheries Research Board of Canada* 36: 716–24

Nemec, Thomas F. 1972. 'I Fish with My Brother: The Structure and Behaviour of Agnatic-Based Fishing Crews in a Newfoundland Irish Outport.' In Andersen and Wadel 1972, 9–34

Netting, R. 1976. 'What Alpine Peasants Have in Common: Observations on Communal Tenure in a Swiss Village.' *Human Ecology* 4: 2, 135–46

Nettler, Gwyn. 1970. *Explanations*. New York: McGraw-Hill

Newfoundland and Labrador. Department of Finance. Economics and Statistics Division. *Historical Statistics of Newfoundland and Labrador*. 4 vols. St John's: Division of Printing Services, Department of Public Works and Services

– Ministry of Fisheries. 1978. *White Paper on Strategies and Programs for Fisheries Development to 1985*. St John's: Government of Newfoundland and Labrador

– Ministry of Fisheries. 1980. *The Fishery: A Business and a Way of Life*. St John's: Government of Newfoundland and Labrador

– Royal Commission on Employment and Unemployment. 1986. *Building on Our Strengths: Report on the Royal Commission on Employment and Unemployment in Newfoundland*. Chaired by J. Douglas House. St John's: Queen's Printer

– Royal Commission to Inquire into the Fishery of Newfoundland and Labrador. 1981. *Report: Phases II and III*. Chaired by Brose Paddock. St John's: The Commission

Newfoundland Fishermen, Food and Allied Workers Union. *See* NFFAWU

NFFAWU. 1977. 'Full-Time Fishermen Get Priority.' *Union Forum* (March 1979), 17. St John's: NFFAWU

– 1979a. 'Licensing: Prompt Action Needed.' *Union Forum* (July 1979), 13–14. St John's: NFFAWU

– 1979b. 'The Need for Licensing Controls.' *Union Forum* (March 1979), 13–14. St John's: NFFAWU

– 1980a. 'Licence the Man, Not the Boat.' *Union Forum* (March 1980), 21–2. St John's: NFFAWU

– 1980b. 'Licensing System Under Review.' *Union Forum* (April 1980), 17–18. St John's: NFFAWU

Novak, Wreslaw S. Wladyslaw. 1980. *'Like It or Not – You Will Be Resettled': Some Economic and Geographical Implications of the Licensing Policy in the Newfoundland Fishery*. Mount Pearl, Nfld: Hawk Duplicating

Oakerson, Ronald J. 1986. 'A Model for the Analysis of Common Property Problems.' In National Research Council (Washington, DC), Office of International Affairs, Board on Science and Technology for International

Development, ed., *Proceedings of the Conference on Common Property Resource Management*, 13–30. Washington, DC: National Academy Press

Oberschall, Anthony, and Eric M. Leifer. 1986. 'Efficiency and Social Institutions: Uses and Misuses of Economic Reasoning in Sociology.' In Ralph H. Turner and James F. Short, Jr., eds., *Annual Review of Sociology* 12: 233–53

Ostrom, Elinor. 1987a. 'Institutional Arrangements for Resolving the Commons Dilemma.' In McCay and Acheson 1987, 250–65

– 1987b. 'Micro-Constitutional Change in a Multi-Constitutional Political System.' Paper presented at the Advances in Comparative Institutional Analysis Conference, Dubrovnik, Yugoslavia, October 19–23

– 1988. *The Commons and Collective Action*. Four lectures delivered in the Program in Political Economy, Harvard University, April 18–21

– 1992. *Crafting Institutions for Self-Governing Irrigation Systems*. San Francisco: ICS (Institute for Contemporary Studies) Press

Paddock Royal Commission. *See* Newfoundland and Labrador. Royal Commission to Inquire into the Inshore Fishery of Newfoundland and Labrador

Paine, Robert. 1971. 'A Theory of Patronage and Brokerage.' In Robert Paine, ed., *Patrons and Brokers in the East Arctic*, 8–21. Newfoundland Social and Economic Papers no. 2. St John's: Institute of Social and Economic Research, Memorial University of Newfoundland

Pearse, Peter H., and James E. Wilen. 1979. 'Impact of Canada's Pacific Salmon Fleet Control Program.' *Journal of the Fisheries Research Board of Canada* 36: 763–81

Perlin, Albert B. 1937. 'An Outline of Newfoundland History.' In Joseph R. Smallwood, ed., *The Book of Newfoundland*, vol. 1. St John's: Newfoundland Book Publishers

Philbrook, Tom. 1965. *Fishermen, Logger, Merchant, Miner: Social Change and Industrialization in Three Newfoundland, Communities*. Newfoundland Social and Economic Studies no. 1. St John's: Institute of Social and Economic Research, Memorial University of Newfoundland

Porter, Marilyn. 1985. 'She Was Skipper of the Shore-Crew': Notes on the History of the Sexual Division of Labour in Newfoundland.' *Labour/Le Travail* 15: 105–23

Prowse, D.W. 1895. *A History of Newfoundland from the English, Colonial and Foreign Records*. London: Macmillan

Retting, R. Bruce, and Jay J.C. Ginter, eds., 1978. *Limited Entry as a Fishery Management Tool: Proceedings of a National Conference to Consider Limited Entry as a Tool in Fishery Management*. Seattle: Institute for Marine Studies, University of Washington

Robinson, W.S. 1950. 'Ecological Correlations and the Behaviour of Individuals.' *American Sociological Review* 15: 351–7

Rogers, J.D. 1911. *Newfoundland*. Vol. 5, Part 4, of *A Historical Geography of the British Colonies*. Oxford: Clarendon Press

Sabetti, Filippo. 1985. 'Community Self-Help and the Law and Regulations of Government.' Paper prepared for the Liberty Fund Seminar, 'The Concept of Community and the Problem of Power,' at Niagara-on-the-Lake, Ont., April 25–8

Schutz, Alfred. 1971a. 'Choosing among Projects of Actions.' In *Collected Papers*. Vol. 1, *The Problem of Social Reality*, 67–96. The Hague: Martinus Nijhoff

– 1971b. 'Common Sense and Scientific Interpretation of Human Action.' In *Collected Papers*. Vol. 1, *The Problem of Social Reality*, 3–47. The Hague: Martinus Nijhoff

Scott, Anthony. 1979. 'Development of Economic Theory on Fisheries Regulation.' *Journal of the Fisheries Research Board of Canada* 36: 725–41

Seary, E.R. 1967. 'The Place Names of Newfoundland.' In J.R. Smallwood, ed., *The Book of Newfoundland* 3: 257–64. St John's: Newfoundland Book Publishers

– 1977. *Family Names of the Island of Newfoundland*. St John's: Memorial University of Newfoundland

Sinclair, Peter R. 1981. 'Return of the Cod: State Intervention in the Newfoundland Fisheries.' Paper presented to the Conference on the Political Economy of Food and Agriculture in Advanced Industrial Societies, Toronto, August

– 1985. *From Traps to Draggers: Domestic Commodity Production in Northeast Newfoundland, 1850–1982*. Newfoundland Social and Economic Studies no. 31. St John's: Institute of Social and Economic Research, Memorial University of Newfoundland

– 1986. 'Saving the Fishery – Again! The 1981–82 Fisheries Crisis in Newfoundland.' In Rex Clark, ed., *Contrary Winds: Essays on Newfoundland Society in Crisis*, 37–55. St John's: Breakwater Books.

– 1987. *State Intervention and the Newfoundland Fisheries*. Aldershot, Hampshire: Avebury – Gower Publishing

Smallwood, Hon. Joseph R. 1975. 'The Brilliant Success Story of Corner Brook.' In Joseph R. Smallwood, ed., *The Book of Newfoundland* 6: 265–75. St John's: Newfoundland Book Publishers (1967)

Stocks, Anthony. 1987. 'Resource Management in an Amazon Varzea Lake Ecosystem.' In McCay and Acheson 1987, 108–20

Swaminathan, M.S. 1986. Foreword to *Proceedings of the Conference on Common Property Resource Management*, edited by the National Research

Council, Office of International Affairs, Board on Science and Technology for International Development. Washington, DC: National Academy Press

Szwed, John. 1966. *Private Cultures and Public Imagery: Interpersonal Relations in a Newfoundland Peasant Society*. Newfoundland Social and Economic Studies no. 2. St John's: Institute of Social and Economic Research, Memorial University of Newfoundland

Taylor, Lawrence. 1987. 'The River Would Run Red With Blood': Community and Common Property in an Irish Fishing Settlement.' In McCay and Acheson 1987, 290–307

Thoms, James R. 1967a. 'The Framework of Our History.' In Joseph R. Smallwood, ed., *The Book of Newfoundland* 3: 528–34. St John's: Newfoundland Book Publishers (1967)

– 1967b. 'The Sawmill Pioneers of Newfoundland.' In Joseph R. Smallwood, ed., *The Book of Newfoundland* 4: 417–29. St John's: Newfoundland Book Publishers (1967)

Tucker, Walter B. 1975. 'Grand Falls: Our Most Prosperous Town.' In Joseph R. Smallwood, ed., *The Book of Newfoundland* 6: 295–9. St John's: Newfoundland Book Publishers (1967)

Turner, Ralph. 1962. 'Role-Taking: Process versus Conformity.' In Arnold Rose, ed., *Human Behaviour and Human Processes: An Interactionist Approach*, 20–40. Boston: Houghton Mifflin

Wilen, James E. 1979. 'Fisherman Behaviour and the Design of Efficient Fisheries Regulation Programs.' *Journal of the Fisheries Research Board of Canada* 36: 855–8

Index

access: and licensing, 53–4, 56–7; limiting of, 50–1; open, and common property, 41, 253 n.5; open, and depletion, 43; open, and right to use versus right to exclude, 74; open, versus regulation of fishing process, 154, 242–3; regulation of, 85–8, 154

Acheson, J.M., 72, 79–81, 221

Andersen, R., 10

anthropology, and fisheries research, 10

Barrett, L.G., 39–40

Berger, P.L., 87

Berkes, F., 90, 245

berths, net and trap: cod-trap, and 'draw' for, 150–1, 153–8, 169–74; as collective private property, 172–3; committee for regulation of, 155; depths of, 155–66; 'draw' for, and differentiation of fishery, 218–9; 'draw' for, and eligibility criteria, 157–8; 'draw' for, and legal action to gain right to enter, 173–4;

exclusion from, 157, 171–4; as family common property, 217–22; local regulation of, 82; salmon-net, and 'draw' for, 151–2, 158; traditional regulation of, 103–4, 124–8, 215–26, 236; types of, 155–66, 170–2

Blomquist, W., 85–8

Blumer, H., 11

'bona fide' fishermen, 51, 61, 179, 254 n.21

Bonavista, 17–19, 192–237, 240

British North America Act (1867), and fisheries policy, 39, 54

Brox, O., 10

Cabot, John, 18

Calvert, Sir George (later Lord Baltimore), 166–7, 257 n.6

capelin seiners, fishermen's opposition to, 159–60

Cartier, Jacques, 166

Charleston, 17–18, 95, 97–119, 138–9, 217–18, 240

Chiaramonte, L., 10